Matrix Canonical Forms

notational skills and proof techniques

S. Gill Williamson

2012: First version
2018: Various corrections
cseweb.ucsd.edu/gill/

Preface

This material is a rewriting of notes handed out by me to beginning graduate students in seminars in combinatorial mathematics (Department of Mathematics, University of California San Diego). Topics covered in this seminar were in algebraic and algorithmic combinatorics. Solid skills in linear and multilinear algebra were required of students in these seminars - especially in algebraic combinatorics. I developed these notes to review the students' undergraduate linear algebra and improve their proof skills. We focused on a careful development of the general matrix canonical forms as a training ground.

I would like to thank Dr. Tony Trojanowski and Professor Darij Grinberg for a careful reading of this material and numerous corrections and helpful suggestions. I would also like to thank Professor Mike Sharpe, UCSD Department of Mathematics, for considerable LaTeX typesetting assistance and for his Linux Libertine font options to the newtxmath package.

S. Gill Williamson, 2012
http://cseweb.ucsd.edu/~gill

CONTENTS

CHAPTER 1

Functions and Permutations

Algebraic terminology

In this first section, we summarize for reference certain basic concepts in algebra. These concepts are useful for the material we develop here and are essential for reading related online sources (e.g., Wikipedia).

REMARK 1.1 (**BASIC SETS AND NOTATION**). We use the notation $\mathbb{N} = \{1, 2, \ldots\}$ for the positive integers. Let $\mathbb{N}_0 = \{0, 1, 2, \ldots\}$ denote the nonnegative integers, and let $\mathbb{Z} = \{0, \pm 1, \pm 2, \ldots\}$ denote the set of all integers. Let $\times^n S$ (n-fold Cartesian product of S) be the set of n-tuples from a nonempty set S. We also use S^n for this cartesian product. A slightly more general notation is to write $S^{\underline{n}}$ for this product where $\underline{n} = \{1, \ldots, n\}$ and the exponential notation R^D denotes all functions 1.38 from D to R. (We use *delta notation*: $\delta(\text{Statement}) = 1$ if Statement is true, 0 if Statement is false.)

Semigroup→Monoid→Group: sets with one binary operation

A function $w : S^2 \longrightarrow S$ is called a *binary operation*. It is sometimes useful to write $w(x, y)$ in a simpler form such as $x\, w\, y$ or simply $x \cdot y$ or even just $x\, y$. To tie the binary operation w to S explicitly, we write (S, w) or $(S, \cdot)$.

DEFINITION 1.2 (**SEMIGROUP**). Let $(S, \cdot)$ be a *nonempty* set S with a binary operation "$\cdot$" . If $(x \cdot y) \cdot z = x \cdot (y \cdot z)$, for all $x,\ y,\ z \in S$, then the binary operation "$\cdot$" is called *associative* and $(S, \cdot)$ is called a *semigroup*. If two elements $s,\ t \in S$ satisfy $s \cdot t = t \cdot s$ then we say s and t *commute*. If for all $x,\ y \in S$ we have $x \cdot y = y \cdot x$ then $(S, \cdot)$ is a *commutative* (or *abelian*) semigroup.

REMARK 1.3 (**SEMIGROUP**). Let $S = \mathbf{M}_{2,2}(\mathbb{Z}_e)$ be the set of 2×2 matrices with entries in $\mathbb{Z}_e = \{0, \pm 2, \pm 4, \ldots\}$, the set of even integers. Define $w(X, Y) = XY$ to be the standard multiplication of matrices (which is associative). Then (S, w) is a semigroup. This semigroup is not commutative (alternatively, it is a noncommutative semigroup or a semigroup with non-commuting elements). The semigroup of even integers, $(\mathbb{Z}_e, \cdot)$, where "$\cdot$" denotes multiplication of integers, is commutative.

Associative + Identity = Monoid

DEFINITION 1.4 (**MONOID**). Let $(S, \cdot)$ be a semigroup. If there exists an element $e \in S$ such that for all $x \in S$, $e \cdot x = x \cdot e = x$, then e is called an *identity* for the semigroup. A semigroup with an identity is called a *monoid.* If $x \in S$ and there is a $y \in S$ such that $x \cdot y = y \cdot x = e$ then y is called an *inverse* of x.

REMARK 1.5 (**MONOID**). The identity is unique (i.e., if e and e' are both identities then $e = e \cdot e' = e'$). Likewise, if y and y' are inverses of x, then $y' = y' \cdot e = y' \cdot (x \cdot y) = (y' \cdot x) \cdot y = e \cdot y = y$ so the inverse of x is unique. Note that this last computation shows that if y' satisfies $y'x = e$ (y' is a "left inverse") and y satisfies $xy = e$ (y is a "right inverse") then $y' = y$. The 2×2 matrices, $\mathbf{M}_{2,2}(\mathbb{Z})$, with matrix multiplication form a monoid (identity I_2, the 2×2 identity matrix).

Associative + Identity + Inverses = Group

DEFINITION 1.6 (**GROUP**). Let $(S, \cdot)$ be a monoid with identity e and let $x \in S$. If there is a $y \in S$ such that $x \cdot y = y \cdot x = e$ then y is called an *inverse* of x (see 1.4). A monoid in which every element has an inverse is a *group.*

REMARK 1.7 (**GROUP**). Commutative groups, $x \cdot y = y \cdot x$ for all x and y, play an important role in group theory. They are also called *abelian* groups. Note that the inverse of an element x in a group is unique: if y and y' are inverses of x, then $y' = y' \cdot e = y' \cdot (x \cdot y) = (y' \cdot x) \cdot y = e \cdot y = y$ (see 1.5).

Ring: one set with two intertwined binary operations

DEFINITION 1.8 (**RING AND FIELD**). A *ring*, $R = (S, +, \cdot)$, is a set with two binary operations such that $(S, +)$ is an abelian group with identity denoted by 0 ("+" is called "addition") and $(S, \cdot)$ is a semigroup ("$\cdot$" is called "multiplication"). The two operations are related by distributive rules which state that for all x, y, z in S:

$$\textbf{(left)}\ \ z \cdot (x + y) = z \cdot x + z \cdot y \ \text{ and } \ (x + y) \cdot z = x \cdot z + y \cdot z \ \ \textbf{(right)}.$$

REMARK 1.9 (**NOTATION, SPECIAL RINGS AND GROUP OF UNITS**). The definition of a ring assumes only that $(S, \cdot)$ is a semigroup (1.2). Thus, a ring may not have a multplicative identity. We refer to such a structure as a *ring without an identity.* Let $R = (S, +, \cdot)$ be a ring. The identity of the abelian group $(S, +)$ is denoted by 0_R (or 0) and is called the *zero* of the ring $(S, +, \cdot)$. If $r \in S$ then the inverse of r in $(S, +)$ is denoted by $-r$ so $r + (-r) = (-r) + r = 0$. Suppose $(S, \cdot)$

is a monoid with identity 1_R (we say "R is a *ring with identity* 1_R"); its invertible elements (or *units*), $U(R)$, form a group, $(U(R), \cdot)$, with 1_R as the group identity. The group $(U(R), \cdot)$, or simply $U(R)$, is the *group of units* of the ring R. If $(S, \cdot)$ is commutative then $(S, +, \cdot)$ is *a commutative ring*. If $(S - \{0\}, \cdot)$ is a group (i.e., $(U(R), \cdot) = (S - \{0\}, \cdot)$) then the ring R is called a *skew-field* or *division ring*. If this group is *abelian* then the ring is called a *field*.

REMARK 1.10 (**BASIC RING IDENTITIES**). If r, s, t are in a ring $(S, +, \cdot)$ then the following basic identities (in braces, plus hints for proof) hold:

(1) $\{r \cdot 0 = 0 \cdot r = 0\}$ If $x + x = x$ then $x = 0$. Take $x = r \cdot 0$ and $x = 0 \cdot r$.

(2) $\{(-r) \cdot s = r \cdot (-s) = -(r \cdot s)\}$ $r \cdot s + (-r) \cdot s = 0 \implies (-r) \cdot s = -(r \cdot s)$.

(3) $\{(-r) \cdot (-s) = r \cdot s\}$ Replace r by $-r$ in **(2)**. Note that $-(-r) = r$.

In particular, if $(S, \cdot)$ has identity $1_R \equiv 1$, then $(-1) \cdot a = -a$ for any $a \in S$ and, taking $a = -1$, $(-1) \cdot (-1) = 1$. It is convenient to define $r - s = r + (-s)$. Then we have $(r - s) \cdot t = r \cdot t - s \cdot t$ and $t \cdot (r - s) = t \cdot r - t \cdot s$:

$$t \cdot (r - s) = t \cdot (r + (-s)) = t \cdot r + t \cdot (-s)) = (t \cdot r + -(t \cdot s)) = t \cdot r - t \cdot s.$$

A field is a ring* $(S, +, \cdot)$ *where* $(S - \{0\}, \cdot)$ *is an abelian group

REMARK 1.11 (**RINGS, FIELDS, IDENTITIES AND UNITS**). The 2×2 matrices over the even integers, $\mathbf{M}_{2,2}(\mathbb{Z}_e)$, with the usual multiplication and addition of matrices, is a noncommutative ring without an identity. The matrices, $\mathbf{M}_{2,2}(\mathbb{Z})$, over all integers, is a noncommutative ring with an identity. The group of units, $U(\mathbf{M}_{2,2}(\mathbb{Z}))$, is all invertible 2×2 integral matrices. The matrix $P = \begin{pmatrix} +1 & -1 \\ -2 & +3 \end{pmatrix}$ is a unit in $\mathbf{M}_{2,2}(\mathbb{Z})$ with $P^{-1} = \begin{pmatrix} 3 & 1 \\ 2 & 1 \end{pmatrix}$. $U(\mathbf{M}_{2,2}(\mathbb{Z}))$ is usually denoted by $\mathrm{GL}(2, \mathbb{Z})$ and is called a *general linear group*. The ring of 2×2 matrices of the form $\begin{pmatrix} x & y \\ -\overline{y} & \overline{x} \end{pmatrix}$ where x and y are complex numbers is a skew-field but not a field. This skew-field is equivalent to (i.e, a "matrix representation of") the *skew field of quaternions* (see Wikipedia article on quaternions). The most important fields for us will be the fields of real and complex numbers.

DEFINITION 1.12 (**IDEAL**). Let $(R, +, \cdot)$ be a ring and let $A \subseteq R$ be a subset of R. If $(A, +)$ is a subgroup of $(R, +)$ then A is a *left ideal* if for every $x \in R$ and $y \in A$, $xy \in A$. A *right ideal* is similarly defined. If A is both a left and right ideal then it is a *two-sided ideal* or, simply, an *ideal*. If $(R, +, \cdot)$ is commutative then all ideals are two-sided.

Remark 1.13 (**Ideal**). The set A of all matrices $a = \begin{pmatrix} x & y \\ 0 & 0 \end{pmatrix}$ forms a subgroup $(A, +)$ of $(\mathbf{M}_{2,2}(\mathbb{Z}), +)$. The subset A is a right ideal but not a left ideal of the ring $\mathbf{M}_{2,2}(\mathbb{Z})$. Note that $R = (A, +, \cdot)$ is itself a ring. This ring has pairs of zero divisors - pairs of elements (a, b) where $a \neq 0$ and $b \neq 0$ such that $a \cdot b = 0$. For example, take the pair (a, b) to be $a = \begin{pmatrix} 0 & 1 \\ 0 & 0 \end{pmatrix}$ and $b = \begin{pmatrix} 1 & 1 \\ 0 & 0 \end{pmatrix}$. The pair (a, b) is, of course, also a pair of zero divisors in $\mathbf{M}_{2,2}(\mathbb{Z})$.

Another example of an ideal is the set of even integers, $\mathbb{Z}_e$, which is a subset of the integers, $\mathbb{Z}$ (which, it is worth noting, forms a ring with *no zero divisor pairs*). The subset $\mathbb{Z}_e$ is an ideal (two-sided) in $\mathbb{Z}$. Given any integer $n \neq 0$, the set $\{k \cdot n \mid k \in \mathbb{Z}\}$ of multiples of n is an ideal of the ring $\mathbb{Z}$ which we denote by $(n) = n\mathbb{Z} = \mathbb{Z}n$. Such an ideal (i.e., generated by a single element, n) in $\mathbb{Z}$ is called a *principal ideal*. It is easy to see that all ideals A in $\mathbb{Z}$ are principal ideals, (n), where $|n| \geq 0$ is minimal over the set A. Another nice property of integers is that they *uniquely factor* into primes (up to order and sign).

Definition 1.14 (**Characteristic of a ring**). Let R be a ring. Given $a \in R$ and an integer $n > 0$, define $na \equiv a + a + \cdots + a$ where there are n terms in the sum. If there is an integer $n > 0$ such that $na = 0$ for all $a \in R$ then the *characteristic* of R is the least such n. If no such n exists, then R has *characteristic zero* (see Wikipedia article "Characteristic (algebra)" for generalizations).

Algebraists have defined several important abstractions of the ring of integers, $\mathbb{Z}$. We next discuss four such abstractions: integral domains, principal ideal domains (PID), unique factorization domains (UFD), and Euclidean domains - each more restrictive than the other.

Euclidean Domain $\Longrightarrow$ ***PID*** $\Longrightarrow$ ***UFD***

Definition 1.15 (**Integral domain**). An *integral domain* is a *commutative* ring with identity, $(R, +, \cdot)$, in which there are no *zero divisor pairs*: pairs of nonzero elements (a, b) where $ab = 0$. (See 1.13 for ring *with* pairs of zero divisors.)

Remark 1.16 (**Divisors, units, associates and primes**). For noncommutative rings, an element a is a left zero divisor if there exists $x \neq 0$ such that $ax = 0$ (right zero divisors similarly defined). Let R be a commutative ring. If $a \neq 0$ and b are elements of a R, we say that a is a *divisor* of b (or *a divides b*), $a \mid b$, if there exists c such that $b = ac$. Otherwise, a does not divide b, $a \nmid b$. Note that if $b = 0$ and $a \neq 0$ then $a \mid b$ because $b = a0$ $(c = 0)$. Thus $a \mid 0$ or a divides 0. (The term *zero divisors* of 1.15 refers to pairs (a, b) of nonzero elements and is not the same as "a is a divisor of 0" or "a divides 0".) An element u in $R - \{0\}$ is an *invertible element* or a *unit of R* if u has an inverse in $(R - \{0\}, \cdot)$. The units form a group, $U(R)$ (1.9). For commutative rings, $ab = u$,

u a unit, implies that both a and b are units: $ab = u$ implies $a(bu^{-1}) = e$ and $(au^{-1})b = e$ so both a and b are units. Two elements a and b of R are *associates in* R if $a = bu$ where u is a unit. An element p in $R - \{0\}$ is *irreducible* if $p = ab$ implies that either a or b is a unit and *prime* if $p \mid ab$ implies $p \mid a$ or $p \mid b$. For unique factorization domains (1.17), p is irreducible if and only if it is prime. In the ring $\mathbb{Z}$, the only invertible elements are $\{+1, -1\}$. The only associates of an integer $n \neq 0$ are $+n$ and $-n$. The integer $12 = 3 \cdot 4$ is the product of two non-units so 12 is not irreducible (i.e., *is* reducible) or, equivalently in this case, not a prime. The integer 13 is a prime with the two associates $+13$ and -13. A field is an integral domain in which every nonzero element is a unit. In a field, if $0 \neq p = ab$ then *both* a and b are nonzero and hence both are units (so at least one is a unit) and thus every nonzero element in a field is irreducible (and prime).

DEFINITION 1.17 (**UNIQUE FACTORIZATION DOMAIN**). An integral domain R is a *unique factorization domain* (UFD) if

(1) Every $0 \neq a \in R$ can be factored into a finite (perhaps empty) product of primes and a unit: $a = up_1 \cdots p_r$. An empty product is defined as 1_R.

(2) If $a = up_1 \cdots p_r$ and $a = wq_1 \cdots q_s$ are two such factorizations then either both products of primes are empty (and $u = w$) or $r = s$ and the q_i can be reindexed so that p_i and q_i are associates for $i = 1, \ldots, s$.

REMARK 1.18 (**UNIQUE FACTORIZATION DOMAINS**). The integers, $\mathbb{Z}$, are a unique factorization domain. Every field is also a unique factorization domain because every nonzero element is a unit times the empty product. If R is a UFD then so are the polynomial rings $R[x]$ and $R[x_1, \ldots, x_n]$. If $a_1, \ldots, a_n$ are nonzero elements of a UFD, then there exists a greatest common divisor $d = \gcd(a_1, \ldots, a_n)$ which is unique up to multiplication by units. The divisor d is *greatest* in the sense that any element $\hat{d}$ such that $\hat{d} \mid a_i$, $i = 1, \ldots, n$, also divides d (i.e., $\hat{d} \mid d$).

DEFINITION 1.19 (**PRINCIPAL IDEAL DOMAIN**). An integral domain R is a *principal ideal domain* (PID) if every ideal in R is a principal ideal (1.13).

REMARK 1.20 (**PRINCIPAL IDEAL DOMAINS**). We noted in Remark 1.13 that every ideal in $\mathbb{Z}$ is a principal ideal. If $(F, +, \cdot)$ is a field, then any ideal $A \neq (0)$ contains a nonzero and hence invertible element a. The ideal $(a) = F$. There is only one nontrivial ideal in a field and that is a principal ideal that equals F. Thus, any field F is a PID. Let $a_1, \ldots, a_n$ be nonzero elements of a PID, R.

It can be shown that if $d = \gcd(a_1, \ldots, a_n)$ in R then there exists $r_1, \ldots, r_n$ in R such that $r_1 a_1 + \cdots + r_n a_n = d$. The ring of polynomials, $F[x_1, \ldots, x_n]$, in $n \geq 2$ variables over field F is not a PID. Also, the ring of polynomials with integral coefficients, $\mathbb{Z}[x]$, is not a PID. For example, the ideal $< 2, x >= \{2a(x)+xb(x) \mid a, b \in \mathbb{Z}[x]\}$ is not a principal ideal $(p(x)), p \in \mathbb{Z}[x]$. Otherwise, $2 = q(x)p(x)$ for some $q \in \mathbb{Z}[x]$ which implies $p = \pm 1$ or $p = \pm 2$, either case leading to a contradiction.

Definition 1.21 (**Euclidean valuation**). Let R be an integral domain and let $v : R - \{0\} \longrightarrow \mathbb{N}_0$ (nonnegative integers). v is a *Euclidean valuation* on R if

(1) For all $a, b \in R$ with $b \neq 0$, there exist q and r in R such that $a = b \cdot q + r$ where either $r = 0$ or $v(r) < v(b)$.

(2) For all $a, b \in R$ with $a \neq 0$ and $b \neq 0$, $v(a) \leq v(a \cdot b)$.

Definition 1.22 (**Euclidean domain**). An integral domain R is a *Euclidean domain* if there exists a Euclidean valuation on R (see 1.21).

Remark 1.23 (**Euclidean domains**). It can be shown that every Euclidian domain is a principal ideal domain and every principal ideal domain is a unique factorization domain. The integers $\mathbb{Z}$ are a Euclidean domain with $v(n) = |n|$. The units of $\mathbb{Z}$ are $\{-1, +1\}$. In general in a Euclidean domain, if a and b are nonzero and b is not a unit then $v(a) < v(ab)$ (check this out for $\mathbb{Z}$). Intuitively, the units have minimal valuations over all elements of the Euclidean domain and multiplying any element by a non-unit increases the valuation. Any field $(F, +, \cdot)$ is a Euclidean domain with $v(x) = 1$ for all nonzero x. The polynomials, $\mathbb{F}[x]$, with coefficients in a field $\mathbb{F}$ form a Euclidean domain with $v(p(x))$ the degree of $p(x)$. The units of $\mathbb{F}[x]$ are all nonzero constant polynomials (degree zero). The ring of polynomials with integral coefficients, $\mathbb{Z}[x]$, is not a PID (1.20) and thus not a Euclidean domain. Likewise, the ring of polynomials in n variables, $n > 1$, over a field $\mathbb{F}$, $\mathbb{F}[x_1, \ldots, x_n]$, is not a PID (1.20) and hence not a Euclidean domain. Rings that are PIDs but not Euclidean domains are rarely discussed (the ring $\mathbb{Z}[\alpha] = \{a + b\alpha \mid a, b \in \mathbb{Z},\ \alpha = (1 + (19)^{1/2} i)\}$ is a standard example).

We now combine a ring with an abelian group to get a module.

Definition 1.24 (**Module**). Let $(R, +, \cdot)$ be a ring with identity 1_R. Let $(M, \oplus)$ be an abelian group. We define an operation with domain $R \times M$ and range M which for each $r \in R$ and $x \in M$ takes (r, x) to rx (juxtaposition of r and

x). This operation, called *scalar multiplication*, defines a *left R-module M* if the following hold for every $r, s \in R$ and $x, y \in M$:

(1) $r(x \oplus y) = rx \oplus ry$ (2) $(r+s)x = rx \oplus sx$ (3) $(r \cdot s)x = r(sx)$ (4) $1_R x = x$.

We sometimes use "+" for the addition in both abelian groups and replace "·" with juxtaposition. Thus, we have: (2) $(r+s)x = rx + sx$ (3) $(rs)x = r(sx)$.

Sometimes a module is defined without the assumption of the identity 1_R. In that case, what we call a module is called a *unitary* module.

Remark 1.25 (**Module**). Let R be the ring of 2×2 matrices over the integers, $\mathbf{M}_{2,2}(\mathbb{Z})$. Let M be the abelian group, $\mathbf{M}_{2,1}(\mathbb{Z})$, of 2×1 matrices under addition. We define juxtaposition to be the usual matrix by column multiplication. Then (1) and (2) correspond to the distributive law for matrix multiplication, (3) is the associative law, and (4) is multiplication on the left by the 2×2 identity matrix. Thus, $\mathbf{M}_{2,1}(\mathbb{Z})$ is an $\mathbf{M}_{2,2}(\mathbb{Z})$-module. If $x \in \mathbf{M}_{2,1}(\mathbb{Z})$ then, obviously, $x + x = x$ implies that $x = \theta_{21}$ where θ_{21} is the zero matrix in $\mathbf{M}_{2,1}(\mathbb{Z})$. To see this, just add $-x$ to both sides of $x + x = x$. This fact is true in any module for the same reason. In particular, in any module if $z \in M$ and $0 \in R$ is the identity of $(R, +)$, then $0z = (0+0)z = 0z + 0z$ and, taking $x = 0z$ in the identity $x + x = x$, $0z = \theta$, the zero in $(M, +)$. This fact is obvious in our $\mathbf{M}_{2,2}(\mathbb{Z})$-module $\mathbf{M}_{2,1}(\mathbb{Z})$. Likewise for any module, if $\alpha \in R$ and θ is the zero in $(M, +)$, then $\alpha\theta = \alpha(\theta + \theta) = \alpha\theta + \alpha\theta$ implies that $\alpha\theta = \theta$. In the modules that we will be interested in (i.e., vector spaces 1.27) it is true that for $\alpha \in R$ and $z \in M$, $\alpha z = \theta$ implies that either $\alpha = 0$ or $z = \theta$. This assertion is not true in our $\mathbf{M}_{2,2}(\mathbb{Z})$-module $\mathbf{M}_{2,1}(\mathbb{Z})$:

$$\alpha z = \begin{pmatrix} 1 & 0 \\ 1 & 0 \end{pmatrix} \begin{pmatrix} 0 \\ 1 \end{pmatrix} = \begin{pmatrix} 0 \\ 0 \end{pmatrix} = \theta_{21}.$$

Remark 1.26 (**Free modules**). Of special interest to us are certain R modules, R^n (see 1.1), where R is a ring with identity 1_R. The abelian groups of these modules consist of n-tuples (n-vectors) of elements in R where addition is component-wise. The multiplication of n-tuples $(x_1, \ldots, x_n)$ by elements α of R is defined component wise: $\alpha(x_1, \ldots, x_n) = (\alpha x_1, \ldots, \alpha x_n)$. Such modules are called *free modules of rank n over R*. For a careful discussion see Wikipedia "Free module."

Definition 1.27 (**Vector space and algebra**). If an abelian group $(M, +)$ is an F-module where F is a field (1.11), then we say $(M, +)$ (or, simply, M) is a *vector space over F* (or M is an *F vector space*). Suppose $(M, +, \cdot)$ is a ring where $(M, +)$ is a vector space over F. Then $(M, +, \cdot)$ is an *algebra over F* (or M is an *F algebra*) if the following scalar rule holds:

scalar rule for all $\alpha \in F$, $a, b \in M$ we have $\alpha(a \cdot b) = (\alpha a) \cdot b = a \cdot (\alpha b)$.

Remark 1.28 (**Vector spaces versus modules**). We will use certain modules over special Euclidean domains (e.g., integers and polynomials) as a tool to understand properties of finite dimensional vector spaces. One basic difference between finite dimensional vector spaces and general modules is that a proper subspace of such a vector space always has lower dimension (rank) than the vector space itself – not so in general for modules. As an example, consider the integers $\mathbb{Z}$. The ordered pairs $\mathbb{Z} \times \mathbb{Z}$ is a $\mathbb{Z}$-module with the usual operations on ordered pairs (free module of rank 2 over $\mathbb{Z}$). The natural "module basis" is $E = \{(0,1),(1,0)\}$ so this module has "rank 2". Take $E' = \{(0,1),(2,0)\}$. The span of E' is a proper submodule of $\mathbb{Z} \times \mathbb{Z}$ over the integers $\mathbb{Z}$ since the first component of every element in the span is even, but the "module rank" of this proper submodule is still 2. If we had used the field of rational numbers $\mathbb{Q}$ instead and regarded E' as a set in the vector space $\mathbb{Q} \times \mathbb{Q}$ over $\mathbb{Q}$ then the span of E' is the entire vector space, not a proper subspace as in the case of $\mathbb{Z}$. We are not defining our terms here, but the basic idea should be clear.

Remark 1.29 (**Complex matrix algebra**). Let $\mathbb{C}$ be the field of complex numbers and let M be $\mathbf{M}_{2,2}(\mathbb{C})$, the additive abelian group of 2×2 matrices with complex entries. Conditions (1) to (4) of 1.24 are familiar properties of multiplying matrices by scalars (complex numbers). Thus, M is a complex vector space or, alternatively, M is a vector space over the field of complex numbers, $\mathbb{C}$. If we regard M as the ring, $\mathbf{M}_{2,2}(\mathbb{C})$, of 2×2 complex matrices using the standard multiplication of matrices, then it follows from the definitions of matrix multiplication and multiplication by scalars that the scalar rule of 1.27 holds, and $\mathbf{M}_{2,2}(\mathbb{C})$ is an algebra over $\mathbb{C}$.

Remark 1.30 (**Rings $\mathbb{K}$ and fields $\mathbb{F}$**). Fields $\mathbb{F}$ of interest will be $\mathbb{Q}$ (rational numbers), $\mathbb{R}$ (real numbers), $\mathbb{C}$ (complex numbers) and the fields of rational functions (ratios of polynomials) $\mathbb{Q}(z)$, $\mathbb{R}(z)$ and $\mathbb{C}(z)$. Thus, $\mathbb{F} \in \{\mathbb{Q}, \mathbb{R}, \mathbb{C}, \mathbb{Q}(z), \mathbb{R}(z), \mathbb{C}(z)\}$. Let $\mathbb{K} \in \{\mathbb{Z}, \mathbb{F}[x], \mathbb{F}\}$ where $\mathbb{Z}$ denotes the integers and $\mathbb{F}[x]$ the polynomials over $\mathbb{F}$. All of these rings are of characteristic zero (Definition 1.14). Note that $\mathbb{K}$ is a Euclidean domain with valuation absolute value (integers), degree (polynomials), or identically 1 for all nonzero elements (field). The fields $\mathbb{Q}(z), \mathbb{R}(z), \mathbb{C}(z), \mathbb{Q}$ are the *quotient fields* for the Euclidean domains $\mathbb{Q}[z], \mathbb{R}[z], \mathbb{C}[z], \mathbb{Z}$, respectively. We will not have much interest in the polynomials $\mathbb{F}[x]$ or rational functions $\mathbb{F}(x)$ where $\mathbb{F} = \mathbb{Q}(z), \mathbb{R}(z), \mathbb{C}(z)$.

Remark 1.31 (**Greatest common divisor, least common multiple**). Suppose d is a common divisor of $a, b \neq 0$ in $\mathbb{K}$. If for all c, $c \mid d$ whenever $c \mid a$ and $c \mid b$, then d is a *greatest common divisor* of a and b (d is a $\gcd(a,b)$). If d is a $\gcd(a,b)$ then td is a $\gcd(ua, vb)$ for any units $t, u, v \in \mathbb{K}$ (i.e., the gcd is defined "up to units"). If $a \neq 0$ we adopt the convention that a is a $\gcd(a, 0)$. An element $c \in \mathbb{K}$ is a *common multiple* of a and b if $a|c$ and $b|c$. If for all $x \in \mathbb{K}$, $a|x$ and $b|x$ implies $c|x$, then c is a *least common multiple* of a and b (c is an lcm(a,b)). The $\mathrm{lcm}(a,b)$ is determined up to units. We also write $\mathrm{lcm(a,b)} = \mathrm{a} \vee \mathrm{b}$ ("join" of a and b) and $\mathrm{gcd(a,b)} = \mathrm{a} \wedge \mathrm{b}$ ("meet" of a and b).

If $\mathbb{K} = \mathbb{F}$ then all nonzero elements are units so d is a $\gcd(a, b)$ for any nonzero d, a, b. Likewise, d is an $\mathrm{lcm}(a, b)$ for any nonzero d, a, b. For $\mathbb{K} = \mathbb{Z}$, the units are $\{+1, -1\}$, and for $\mathbb{K} = \mathbb{F}[x]$, the units are the nonzero constant polynomials. Thus, we focus on the cases $\mathbb{K} = \mathbb{Z}$ or $\mathbb{F}[x]$: Suppose $a = p_1^{e_1} \cdots p_m^{e_m}$ and $b = p_1^{f_1} \cdots p_m^{f_m}$ are prime factorizations of a and b, and suppose that a and b are positive integers (if $\mathbb{K} = \mathbb{Z}$) or monic polynomials (if $\mathbb{K} = \mathbb{F}[x]$). Then $p_1^{\min(e_1, f_1)} \cdots p_m^{\min(e_m, f_m)}$ is a gcd(a, b), and $p_1^{\max(e_1, f_1)} \cdots p_m^{\max(e_m, f_m)}$ is an lcm(a, b). Note that $ab = \gcd(a, b)\mathrm{lcm}(a, b) = (a \wedge b)(a \vee b)$. If $\mathbb{K} = \mathbb{Z}$ then let $d > 0$ be the largest positive divisor of a and b. The notation $d = \gcd(a, b)$ is sometimes used to indicate that this is the canonical choice up to units for a gcd of the integers a and b. For $\mathbb{K} = \mathbb{F}[x]$ the canonical choice up to units is often taken to be the monic polynomial d (coefficient of highest power one).

Sets, lists, multisets and functions

We consider collections of objects where order and repetition play different roles. The concept of a *function* and the terminology for specifying various types of functions are discussed.

Remark 1.32 (**Notation for sets, lists and multisets**). The empty set is denoted by $\emptyset$. Sets are specified by braces: $A = \{1\}$, $B = \{1, 2\}$. They are unordered, so $B = \{1, 2\} = \{2, 1\}$. Sets $C = D$ if $x \in C$ implies $x \in D$ (equivalently, $C \subseteq D$) and $x \in D$ implies $x \in C$. If you write $C = \{1, 1, 2\}$ and $D = \{1, 2\}$ then, by the definition of set equality, $C = D$. A *list, vector* or *sequence* (specified by parentheses) is ordered: $C' = (1, 1, 2)$ is not the same as $(1, 2, 1)$ or $(1, 2)$. Two lists (vectors, sequences), $(x_1, x_2, \ldots, x_n) = (y_1, y_2, \ldots, y_m)$, are equal if and only if $n = m$ and $x_i = y_i$ for $i = 1, \ldots, n$. A list such as $(x_1, x_2, \ldots, x_n)$ is also written $x_1, x_2, \ldots, x_n$, without the parentheses. There are occasions where we discuss collections of objects where, like sets, order doesn't matter but, like lists, repetitions do matter. These objects are called *multisets*. A multiset can be specified by giving the elements of the multiset with each repeated a certain number of times. For example, $\{1, 1, 2, 2, 2, 3, 3, 3\}$ is a multiset with 1 twice, 2 three, and 3 three times. In this case, $\{1, 1, 2, 2, 2, 3, 3, 3\} \neq \{1, 2, 3\}$ but $\{1, 1, 2, 2, 2, 3, 3, 3\} = \{1, 2, 1, 3, 2, 3, 3, 2\}$. We say 2 is a member of $\{1, 1, 2, 2, 2, 3, 3, 3\}$ with repetition (multiplicity, repetition number) 3. The size of the multiset $\{1, 1, 2, 2, 2, 3, 3, 3\}$ is $2 + 3 + 3 = 8$ (sum of the distinct repetition numbers). The use of braces to define multisets is like the use of braces to define sets. You must make clear in any discussion whether you are discussing sets or multisets. The union of two multisets combines their elements and their multiplicities: $\{1, 1, 2, 2, 2, 3\} \cup \{1, 2, 2, 3, 3\} = \{1, 1, 1, 2, 2, 2, 2, 2, 3, 3, 3\}$.

If S is a finite set, then $|S|$ is the number of elements in S. Thus, $|\{1, 1, 2\}| = |\{1, 2\}| = 2$. $\mathbb{P}(S)$ is the set of all subsets of S, and $\mathbb{P}_k(S)$ is all subsets of S of size k. If $|S| = n$ then $|\mathbb{P}(S)| = 2^n$ and $|\mathbb{P}_k(S)| = \binom{n}{k}$ (binomial coefficient). We use *underline* notation, $\underline{n} = \{1, 2, \ldots, n\}$. We sometimes leave off the underline when the meaning is clear: $\mathbb{P}_k(\underline{n}) = \mathbb{P}_k(n)$.

Definition 1.33 (**Partition of a set**). A *partition* of a set S is a set or *collection*, $\mathcal{B}(S)$, of nonempty subsets, X, of S such that each element of S is contained in *exactly one* set $X \in \mathcal{B}(S)$. The sets $X \in \mathcal{B}(S)$ are called the *blocks* of the partition $\mathcal{B}(S)$. A set $D \subseteq S$ consisting of exactly one element from each block is called a *system of distinct representatives* (or "SDR") for the partition.

Remark 1.34 (**Partition examples**). $\mathcal{B}(S) = \{\{a, c\}, \{b, d, h\}, \{e, f, g\}\}$ is a partition of the set $S = \{a, b, c, d, e, f, g, h\}$. The set $\{c, d, g\}$ is an SDR for this partition. Note that $\{\{a, c\}, \{a, c\}, \{b, d, h\}, \{e, f, g\}\}$ is a partition of S and is the same as $\{\{a, c\}, \{b, d, h\}, \{e, f, g\}\}$. (Recall that repeated elements in a description of a set count as just one element.)

Remark 1.35 (**Equivalence relations**). Consider the partition of 1.34. For each pair $(x, y) \in S \times S$, write $x \sim y$ if x "is in the same block as" y and $x \nsim y$ if x "is not in the same block as" y. For all $x, y, z \in S$ we have

$$(1.36)\qquad (1)\; x \sim x \quad (2)\; x \sim y \implies y \sim x \quad (3)\; x \sim y \text{ and } y \sim z \implies x \sim z.$$

Condition (1) is called *reflexive*, (2) is called *symmetric*, and (3) is called *transitive*. Any relation defined for all $(x, y) \in S \times S$ which can be either true or false can be written as $x \sim y$ if true and $x \nsim y$ if false. If such a relation is reflexive, symmetric and transitive then it is called an *equivalence relation*. Every equivalence relation can be thought of as a partition of S. As an example, take the "is in the same block as" equivalence relation. Suppose we just knew how to check if two things were in the same block but didn't know the blocks. We could reconstruct the set of blocks (i.e., the partition) by taking each $x \in S$ and constructing the set $E_x = \{y \mid x \sim y\}$. This block is called the "equivalence class" of x. The partition could be reconstructed as the set of equivalence classes:

$$(1.37)\qquad \mathcal{B}(S) = \{E_x \mid x \in S\} = \{E_a, E_b, E_c, E_d, E_e, E_f, E_g, E_h\}.$$

In this list we have duplicate blocks (e.g., $E_a = E_c$). But duplicates count as the same element in set notation: $\{1, 1, 2\} = \{1, 2\}$ (1.32). You should carry out the construction and proof of 1.37 for the general equivalence class. You will need to use Definition 1.33. Wikipedia has a good article.

Definition 1.38 (**Functions**). Let A and B be sets. A *function* f from A to B is a rule that assigns to each element $x \in A$ a unique element $y \in B$. We write $y = f(x)$. Two functions f and g from A to B are equal if $f(x) = g(x)$ for all $x \in A$.

This definition is informal as it uses "rule," "assign" and "unique" intuitively, but that is good enough for us as we shall give many examples. Given a function

f from A to B, we can define a set $F \subseteq A \times B$ by

(1.39) $$F = \{(x, f(x)) \mid x \in A\}.$$

We call F the *graph* of f, denoted by Graph(f). A subset $F \subseteq A \times B$ is the graph of a function from A to B if and only if it satisfies the following two conditions:

(1.40) $$\text{G1}: \quad (x, y) \in F \quad \text{and} \quad (x, y') \in F \implies y = y'$$

(1.41) $$\text{G2}: \quad \{x \mid (x, y) \in F\} = A.$$

Condition G1 makes the idea of "unique" more precise, and G2 specifies what is meant by "assigns to each." Two functions, f and g, are equal if and only if their graphs are equal as sets: Graph(f) = Graph(g). The set A is called the *domain* of f (written A = domain(f)), and B is called the *range* of f (written B = range(f)). The notation $f : A \to B$ is used to denote that f is a function with domain A and range B. For $S \subseteq A$, define $f(S)$ (image of S under f) by $f(S) \equiv \{f(x) \mid x \in S\}$. In particular, $f(A)$ is called the *image* of f (written $f(A)$ = image(f)). The set of *all* functions with domain A and range B can be written $\{f \mid f : A \to B\}$ or simply as B^A. If A and B are finite then $|A^B|$ is $|A|^{|B|}$. The *characteristic or indicator function* of a set $S \subseteq A$, $\mathcal{X}_S : A \to \{0, 1\}$, is defined by

(1.42) $$\mathcal{X}_S(x) = 1 \text{ if and only if } x \in S.$$

The *restriction* f_S of $f : A \to B$ to a subset $S \subseteq A$ is defined by

(1.43) $$f_S : S \to B \text{ where } f_S(x) = f(x) \text{ for all } x \in S.$$

If $f : A \to B$ and $g : B \to C$ then the *composition* of g and f, denoted by $gf : A \to C$, is defined by

(1.44) $$gf(x) = g(f(x)) \text{ for } x \in A.$$

Note that composition of functions is associative: $h(gf) = (hg)f$ if $f : A \to B$, $g : B \to C$ and $h : C \to D$. In some discussions, the product of functions, also denoted gf, is defined by $gf(x) = g(x)f(x)$. Another notation for composition of functions is $g \circ f$. Thus, $g \circ f(x) \equiv g(f(x))$.

In the following six examples (1.45), the sets are specified by listing vertically their elements. Arrows collectively define the rule. In the first example, $x = 3$ is in $A = \{1, 2, 3, 4\}$, and $f(x) = a$ is defined by the arrow from 3 to a.

(1.45) **Examples of Functions**

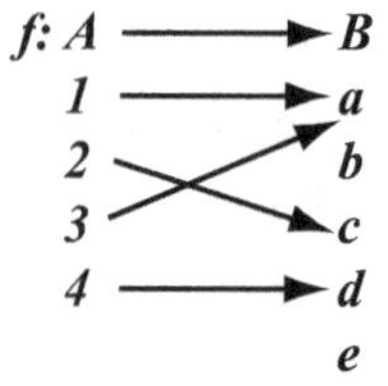

(1) Here, the *domain* of the function is A, the *range* is B, and the *image* is $\{a, c, d\}$. Arrows indicate the element of B that is assigned to a particular element of A.

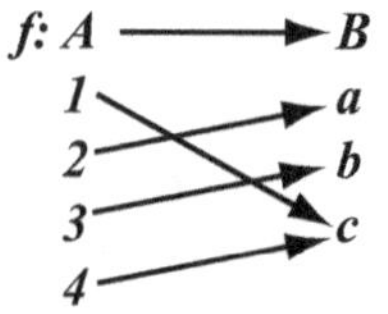

(2) This function is *surjective* or *onto*, its image is $\{a, b, c\}$ which equals its range. To be precise, if y is an element of B then there exists an element x of A such that $f(x) = y$.

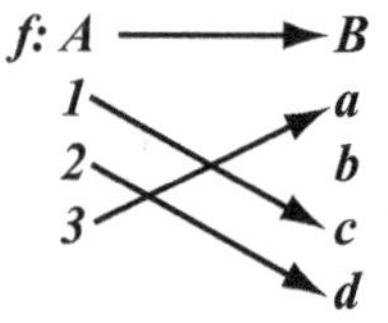

(3) This function is *injective* or *one-to-one*. Its image is $\{a, c, d\}$ which has the same size as its domain. No two elements of its domain are assigned to the same element of its range. If x, y are two different elements of A, then $f(x)$ and $f(y)$ are different elements of B.

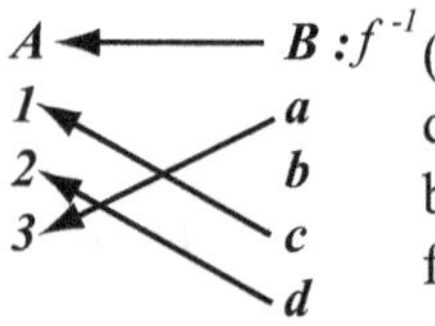

(4) Every injective function f has an *inverse* function denoted by f^{-1} which has domain the image of f and has both range and image equal to the domain of f. The function shown here is the inverse of f in the preceding example.

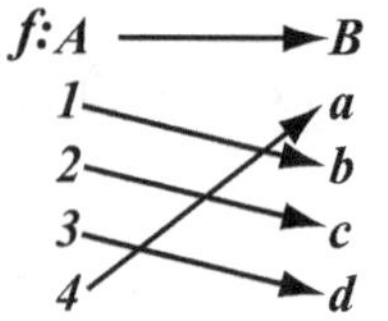

(5) A function that is both injective and surjective is called *bijective*. For finite sets, A and B, there exists a bijection with domain A and range B if and only if A and B have the same size (or cardinality).

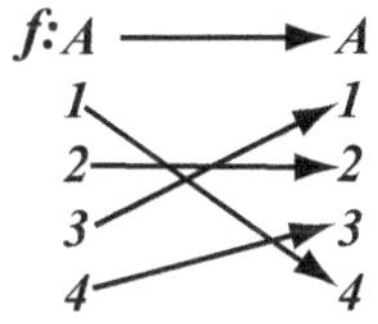

(6) If f is a bijection with domain and range the same, say A, then f is called a *permutation of A* The set of all permutations of A is designated by PER(A) or S(A).

There are many ways to describe a function. Any such description must specify the domain, the range, and the rule for assigning some range element to each domain element. From the discussion following Definition 1.38, you could specify the first function in the above examples (1.45) using set notation: the domain is the set $\{1, 2, 3, 4\}$, the range to be the set $\{a, b, c, d, e\}$, and the function f is the set: $\text{Graph}(f) = \{(1, a), (2, c), (3, a), (4, d)\}$. Alternatively, you could describe the same function by giving the range as $\{a, b, c, d, e\}$ and using *two line notation*

(1.46)
$$f = \begin{pmatrix} 1 & 2 & 3 & 4 \\ a & c & a & d \end{pmatrix}.$$

If we assume the domain, in order, is 1 2 3 4, then 1.46 can be abbreviated to *one line*: $a\,c\,a\,d$.

Definition 1.47 (**Coimage partition**). Let $f : A \to B$ be a function with domain A and range B. Let $\text{image}(f) = \{f(x) \mid x \in A\}$ (1.38). The inverse image of an element $y \in B$ is the set $f^{-1}(y) \equiv \{x \mid f(x) = y\}$. The *coimage* of f is the set of subsets of A :

(1.48)
$$\text{coimage}(f) = \{f^{-1}(y) \mid y \in \text{image}(f)\}.$$

The coimage(f) is a partition of A (1.33) called the *coimage partition of A induced by f*.

For the function f of 1.46, we have $\text{image}(f) = \{a, c, d\}$. Thus, the coimage of f is

(1.49)
$$\text{coimage}(f) = \{f^{-1}(a), f^{-1}(c), f^{-1}(d)\} = \{\{1, 3\}, \{2\}, \{4\}\}.$$

Definition 1.50 (**Sets of functions**). Let $\underline{n} = \{1, 2, \ldots, n\}$ and let $\underline{p}^{\underline{n}}$ be all functions with domain $\underline{n}$, range $\underline{p}$. Define

$$\text{SNC}(n, p) = \{f \mid f \in \underline{p}^{\underline{n}}, i < j \implies f(i) < f(j)\} \text{ (strictly increasing)}$$
$$\text{WNC}(n, p) = \{f \mid f \in \underline{p}^{\underline{n}}, i < j \implies f(i) \leq f(j)\} \text{ (weakly increasing)}$$
$$\text{INJ}(n, p) = \{f \mid f \in \underline{p}^{\underline{n}}, i \neq j \implies f(i) \neq f(j)\} \text{ (injective)}$$
$$\text{PER}(n) = \text{INJ}(n, n) \text{ (permutations of } \underline{n}\text{)}.$$

From combinatorics, $|\text{INJ}(n,p)| = (p)_n = p(p-1)\cdots(p-n+1)$, $|\text{PER(n)}| = \text{n!}$,

$$|\text{SNC}(n,p)| = \binom{p}{n} \quad \text{and} \quad |\text{WNC}(n,p)| = \binom{p+n-1}{n}.$$

More generally, if $X \subseteq \underline{n}$ and $Y \subseteq \underline{p}$, then $\text{SNC}(X,Y)$ denotes the strictly increasing functions from X to Y. We define $\text{WNC}(X,Y)$ and $\text{INJ}(X,Y)$ similarly.

Sometimes "increasing" is used instead of "strictly increasing" or "nondecreasing" instead of "weakly increasing" for the functions of 1.50.

The next identity expresses the set $\text{INJ}(n,p)$ as a composition of functions in $\text{SNC}(n,p)$ and $\text{PER}(n)$. This is illustrated in the table (1.52) that follows for $n = 3, p = 4$ (f, g and the table entries, fg, are in one line notation):

(1.51) $$\text{INJ}(n,p) = \{fg \mid f \in \text{SNC}(n,p),\ g \in \text{PER}(n)\}.$$

(1.52) **Table for** $\text{INJ}(3,4) = \{fg \mid f \in \text{SNC}(3,4),\ g \in \text{PER}(3)\}$

$$f = \begin{pmatrix} 1 & 2 & 3 \\ 2 & 3 & 4 \end{pmatrix} \quad g = \begin{pmatrix} 1 & 2 & 3 \\ 3 & 2 & 1 \end{pmatrix} \quad fg = \begin{pmatrix} 1 & 2 & 3 \\ 4 & 3 & 2 \end{pmatrix}$$

f \ g	1 2 3	1 3 2	2 1 3	2 3 1	3 1 2	3 2 1
1 2 3	1 2 3	1 3 2	2 1 3	2 3 1	3 1 2	3 2 1
1 2 4	1 2 4	1 4 2	2 1 4	2 4 1	4 1 2	4 2 1
1 3 4	1 3 4	1 4 3	3 1 4	3 4 1	4 1 3	4 3 1
2 3 4	2 3 4	2 4 3	3 2 4	3 4 2	4 2 3	4 3 2

REMARK 1.53 (**Example of an** $\text{INJ}(n,p)$ **composition**). Consider $h \in \text{INJ}(3,5)$ where $h = \begin{pmatrix} 1 & 2 & 3 \\ 4 & 1 & 3 \end{pmatrix}$. Following 1.51 and 1.52, we want to write h as the composition fg where $f \in \text{SNC}(3,5)$ and $g \in \text{PER}(3)$. Obviously, $\text{image}(h) = \text{image}(f) = \{1,3,4\}$ (see discussion following 1.38), and thus f is uniquely defined: $f = \begin{pmatrix} 1 & 2 & 3 \\ 1 & 3 & 4 \end{pmatrix}$. To write $h = fg$ we only need the permutation $g = f^{-1}h$ where $f^{-1} = \begin{pmatrix} 1 & 3 & 4 \\ 1 & 2 & 3 \end{pmatrix}$. Thus, $g = \begin{pmatrix} 1 & 2 & 3 \\ 3 & 1 & 2 \end{pmatrix}$.

Permutations

(1.54) **Two line, one line and cycle notation**

There are three standard notations we shall use for writing permutations. The permutation f specified in the sixth arrow diagram of 1.45 can be written in two line notation as

$$f = \begin{pmatrix} 1 & 2 & 3 & 4 \\ 4 & 2 & 1 & 3 \end{pmatrix}.$$

The same permutation could be written as

$$f = \begin{pmatrix} 4 & 3 & 2 & 1 \\ 3 & 1 & 2 & 4 \end{pmatrix}$$

since the same rule of assignment is specified. In the first of the above examples, the second line is $(4, 2, 1, 3)$. Since we know this is the second line of a two line notation for that permutation, we know that $D = R = \{4, 2, 3, 1\} = \{1, 2, 3, 4\}$. In some discussions, the permutations are referred to by only the second line in the two line notation. This shorthand representation of permutations is called the *one line* notation. When using one line notation, the order of the elements in the missing first line (read left to right) must be specified in advance. If that is done, then the missing first line can be used to construct the two line notation and thus the rule of assignment. This seems trivially obvious, but sometimes the elements being permuted have varied and complex natural linear orders. In our example f, above, there are just two natural orders on the set D, increasing and decreasing. If the agreed order is increasing then $f = (4, 2, 1, 3)$ is the correct one line notation. If the agreed order is decreasing then $f = (3, 1, 2, 4)$ is the correct one line notation.

Our third notation for permutations, *cycle* notation, is more interesting. In cycle notation, the permutation f of the previous paragraph is written

(1.55) $$f = (1, 4, 3)(2)$$

The "cycle" $(1, 4, 3)$ is read as "1 is assigned to 4, 4 is assigned to 3, and 3 is assigned to 1." This cycle has three *elements*, 1, 4, and 3, and thus it has *length 3*, the number of elements in the cycle. The cycle (2) has length one and is read as "2 is assigned to 2". The usual convention with cycle notation is to leave out cycles of length one. Thus, we would write

$$f = (1, 4, 3)$$

and we would specify that the set D being permuted is $D = \{1, 2, 3, 4\}$. This latter information is required so we can reconstruct the missing cycles of length one.

(1.56) **Product, inverse and identity**

If f and g are permutations, then the product, $h = fg$, of f and g is their *composition*: $h(x) = (fg)(x) = f(g(x))$. For example, take $f = (1, 4, 3)$ and $g = (1, 5, 6, 2)$ as permutations (cycle notation) of $\underline{6} = \{1, 2, \ldots, 6\}$. Take $x = 5$ and compute $g(5) = 6$ and $f(6) = 6$. Thus, $(fg)(5) = 6$. Continuing in this manner we get

(1.57) $$h = fg = \begin{pmatrix} 1 & 2 & 3 & 4 & 5 & 6 \\ 5 & 4 & 1 & 3 & 6 & 2 \end{pmatrix} \quad \text{or} \quad fg = (1, 5, 6, 2, 4, 3).$$

In computing fg, we mixed cycle notation and two line notation. You should compose $f = (1,4,3)$ and $g = (1,5,6,2)$ to get $h = fg = (1,5,6,2,4,3)$, working entirely in cycle notation. Then put both f and g into two line notation and compose them to get the first identity in 1.57.

The identity permutation, e, on D is defined by $e(x) = x$ for all $x \in D$. For $D = \underline{6}$ we have

(1.58) $$e = \begin{pmatrix} 1 & 2 & 3 & 4 & 5 & 6 \\ 1 & 2 & 3 & 4 & 5 & 6 \end{pmatrix} \quad \text{or} \quad e = (1)(2)\cdots(6)\,.$$

For obvious reasons, we ignore the convention of leaving out cycles of length one when writing the identity permutation in cycle notation.

Every permutation, h, has an inverse, h^{-1}, for which $hh^{-1} = h^{-1}h = e$. For example, if we take $h = fg$ of 1.57, then

(1.59) $$h^{-1} = \begin{pmatrix} 5 & 4 & 1 & 3 & 6 & 2 \\ 1 & 2 & 3 & 4 & 5 & 6 \end{pmatrix} \quad \text{or} \quad h^{-1} = (3,4,2,6,5,1)\,.$$

Note that the two line representation of $h = fg$ in 1.57 was just "flipped over" to get the two line representation of h^{-1} in 1.59. It follows that $(h^{-1})^{-1} = h$ (which is true in general). The cycle representation of $h = fg$ in 1.57 was written in reverse order to get the cycle representation of h^{-1} in 1.59.

(1.60) **Properties of cycles and transpositions**

Two cycles are *disjoint* if they have no entries in common (i.e., are disjoint as sets). Thus, the permutation (2517)(346) of $\underline{7}$ is written as the product of two disjoint cycles. We leave out the commas in cycle notation, writing (2517) rather than $(2,5,1,7)$, when the meaning is clear.

When a permutation is the product of disjoint cycles $f = c_1c_2\ldots c_t$, these cycles can be reordered in any manner (e.g., $f = (12)(34)(56) = (34)(12)(56) = (56)(12)(34)$, etc.) without changing f. Also, the order of the entries in a cycle can be shifted around cyclically without changing the permutation:

$$(2517)(346) = (5172)(346) = (1725)(346) = (7251)(463) = (2517)(634).$$

A cycle of length two, like (27), is called a *transposition*. A cycle such as $c = (1423)$ can be written as a product of transpositions in a number of ways:

(1.61) $$c = (13)(12)(14) = (13)(12)(14)(43)(42)(14)(23)(14)(24).$$

The arguments to the permutations are on the right. Thus, the function c evaluated at the integer 4 is $c[4] = (13)(12)(14)[4] = (13)(12)[1] = (13)[2] = 2$. Note that the number of transpositions in the first "transposition product" representation of c above (1.61) is 3 and in the second representation is 9. Although a given cycle c can be written as a product of transpositions in different ways, say $t_1t_2\cdots t_p$ and $s_1s_2\cdots s_q$, the number of transpositions, p and q, are always both even or both odd (they have the same *parity* or, equivalently,

$p \equiv q \pmod 2$). In the example 1.61, we have $p = 3$ and $q = 9$, both odd. We will discuss this further below.

Given a cycle $c = (x_1 x_2 \cdots x_k)$ of length $|c| = k$, c can always be written as the product of $k - 1$ transpositions. For example,

$$c = (x_1 x_2 \cdots x_k) = (x_1, x_k)(x_1, x_{k-1}) \cdots (x_1, x_3)(x_1, x_2). \tag{1.62}$$

In fact, c cannot be written as a product of less than $k-1$ transpositions.

Definition 1.63 **(Index of permutation)**. Let $f = c_1 c_2 \cdots c_p$ be a permutation written as a product of disjoint cycles c_t, where $|c_t| = k_t$, $t = 1, 2, \ldots, p$. We define $I(f)$, the *index* of f, by

$$\textbf{(Index)}\ \ I(f) = \sum_{t=1}^{p} (k_t - 1). \tag{1.64}$$

It is easy to see that the index, $I(f)$, is the smallest number of transpositions in any transposition product representation of f. As another example, take $f = c_1 c_2 = (1234)(567)$ to be a permutation on $\underline{9}$. The index, $I(f) = 5$. Let's check what happens to the index when we multiply f by a transposition $\tau = (a, b)$, depending on the choice of a and b.

If $\tau = (a, b) = (89)$, then $\tau f = (89)(1234)(567)$ and $I(\tau f) = I(f) + 1$. In this case, neither a nor b is in either cycle c_1 or c_2. If $f = c_1 c_2 \cdots c_k$ where c_i are the disjoint cycles ($|c_i| > 1$) and if a and b are not in any of the c_i, then $\tau f = (a\,b) f$ satisfies $I(\tau f) = I(f) + 1$:

$$\begin{gathered} \textbf{a not in, b not in} : \tau f = (a\,b) c_1 c_2 \cdots c_k\,. \\ I(\tau f) = I(f) + 1 \end{gathered} \tag{1.65}$$

Let $\tau = (a\,b) = (59)$ with $a = 5$, and let $f = (1234)(567)$ be a permutation on $\underline{9}$. In this case, $a = 5$ is in a cycle (the cycle (567)), but b is not in any cycle. Since the cycles commute, let's put the cycle containing a first, $f = (567)(1234)$ (this is just a notational convenience). We compute $\tau f = (59)(567)(1234) = (5679)(1234)$. Applying Definition 1.63, we get $I(\tau f) = I(f) + 1$.

The general situation is to take $f = c_1 c_2 \cdots c_k$ ($|c_i| > 1$), and take $\tau = (a, b)$ where a is in a cycle but b is not. We assume, without loss of generality, that a is in $c_1 = (a\ x_1\ x_2\ \ldots x_r)$. We compute, $(a\,b)(a\ x_1\ \ldots x_r) = (a\ x_1\ \ldots x_r\ b)$. Thus, $I(\tau f) = r + 1 + I(c_2 \cdots c_k)$, $I(f) = r + I(c_2 \cdots c_k)$ and $I(\tau f) = I(f) + 1$. To summarize:

$$\begin{gathered} \textbf{a in, b not in} : (a\,b)(a\ x_1\ \ldots x_r) = (a\ x_1\ \ldots x_r\ b). \\ I(\tau f) = I(f) + 1 \end{gathered} \tag{1.66}$$

Next we consider the case where $f = c_1 c_2 \cdots c_k$ ($|c_i| > 1$), and $\tau = (a, b)$ where a is in a cycle and b is in the same cycle. We assume, without loss of generality, that a and b are in $c_1 = (a\ x_1\ \ldots x_r\ b\ y_1 \ldots y_s)$. We compute,

$(a\,b)(a\;x_1\;\ldots x_r\;b\,y_1\ldots y_s) \;=\; (a\;x_1\;\ldots x_r)\,(b\,y_1\ldots y_s)$. Thus $I(\tau f) = r + s + I(c_2\cdots c_k)$, $I(f) = r + s + 1 + I(c_2\cdots c_k)$ and $I(\tau f) = I(f) - 1$. To summarize,

(1.67) **a, b same** : $(a\,b)(a\;x_1\;\ldots x_r\;b\,y_1\ldots y_s) \;=\; (a\;x_1\;\ldots x_r)\,(b\,y_1\ldots y_s)$.

$$I(\tau f) = I(f) - 1$$

Finally, we consider the reverse of equation 1.67 where a and b are in different cycles. We compute, $(a\,b)(a\;x_1\;\ldots x_r)\,(b\,y_1\ldots y_s) \;=\; (a\;x_1\;\ldots x_r\;b\,y_1\ldots y_s)$. Thus, $I(\tau f) = r{+}s{+}1{+}I(c_2\cdots c_k)$, $I(f) = r{+}s{+}I(c_2\cdots c_k)$ and $I(\tau f) = I(f){+}1$. To summarize,

(1.68) **a, b diff** : $(a\,b)(a\;x_1\;\ldots x_r)\,(b\,y_1\ldots y_s) \;=\; (a\;x_1\;\ldots x_r\;b\,y_1\ldots y_s)$.

$$I(\tau f) = I(f) + 1$$

DEFINITION 1.69 (**PARITY**). We say that m and n in $\mathbb{Z} = \{0, \pm 1, \pm 2, \ldots\}$ have the same *parity* if $m - n$ is even. Equivalently, we can write $m \equiv n \pmod 2$ or $(-1)^m = (-1)^n$.

Recall the index function, $I(f)$, of Definition 1.63.

LEMMA 1.70 (**PARITY OF INDEX AND TRANSPOSITION COUNT**). *Let f be a permutation. Suppose that $f = \tau_1\tau_2\cdots\tau_q$ is any representation of f as a product of transpositions τ_i. Then the parity of q is the same as the parity of the index, $I(f)$, of f.*

PROOF. Note that

(1.71) $$e = (\tau_q\cdots\tau_2\tau_1)(\tau_1\tau_2\cdots\tau_q) = \tau_q\cdots\tau_2\tau_1 f.$$

Equations 1.65 through 1.68 state that multiplying an arbitrary permutation by a transposition τ either increases or decreases that permutation's index by one. Thus, $I(\tau_1 f) = I(f) \pm 1$, $I(\tau_2(\tau_1 f)) = I(f) \pm 1 \pm 1$, etc. Applying this observation to 1.71 inductively gives

$$0 = I(e) = I(f) + p - n$$

where p is the number of times a transposition in the sequence $(\tau_q, \ldots, \tau_2, \tau_1)$ results in a "+1" and n the number of times a transposition results in a "-1". Thus, $I(f) = n - p$. But, $n - p \equiv n + p \pmod 2$. Thus, $I(f) \equiv n + p \pmod 2$. But $n + p = q$, the number of transpositions. This completes the proof that $I(f) \equiv q \pmod 2$ or, equivalently, $I(f)$ and q have the same parity. □

DEFINITION 1.72 (**Sign of a permutation**). Let f be a permutation. The *sign* of f is defined as $\mathrm{sgn}(f) = (-1)^{I(f)}$.

From Lemma 1.70, an alternative definition is $\operatorname{sgn}(f) = (-1)^q$ where q is the number of transpositions in any representation of f as a product of transpositions: $f = \tau_1\tau_2\cdots\tau_q$. As an example, consider the permutation $c = (1423)$ of 1.61.

(1.73) $$c = (1423) = (13)(12)(14)(43)(42)(14)(23)(14)(24).$$

The index, $I(c)$ is 3 so $\operatorname{sgn}(c) = (-1)^3 = -1$. The number of transpositions in the product of transpostions in 1.73 is 9. Thus, $\operatorname{sgn}(c) = (-1)^9 = -1$.

Definition 1.74 **(Inversions of a permutation).** Let S be a set, $|S| = n$, for which there is an agreed upon ordering $s_1 < s_2 < \cdots < s_n$ of the elements. Let f be a permutation of S. An *inversion* of f with respect to this ordering is a pair of elements, $(x, y) \in S \times S$ where $x < y$ but $f(x) > f(y)$. Let $\operatorname{Inv}(f)$ denote the set of all inversions of f with respect to the specified ordering. An inversion of the form (s_i, s_{i+1}), $1 \le i < n$, is called an *adjacent* inversion (s_i and s_{i+1} are "adjacent" or "next to each other" in the ordering on S).

For notational convenience, we take $S = \underline{n}$ with the usual order on integers $(1 < 2 < \cdots < n)$. Figure 1.75 gives an example of a permutation f which is given in two line notation and also in one line notation (at the base of the *inversion grid* used to plot the set $\operatorname{Inv}(f)$).

(1.75) **Figure : Inversion grid**

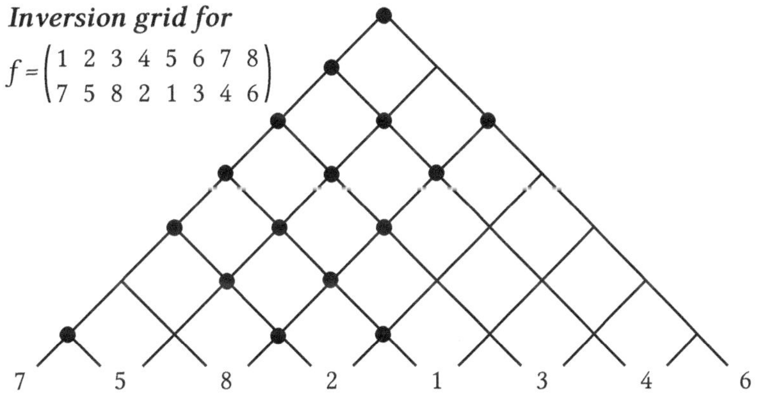

Interchange the 7 and 5 for the inversion (1,2). **A dot for (7,5) disappears. The remaining dots of type (7, x) move down one step to the right. The dots of type (5, x) move up one step to left.**

Each intersection of two lines in the inversion grid (1.75) corresponds to a pair of integers in the one line notation for f. The intersection points corresponding to inversions are marked with solid black dots. There are 16 solid black dots, so $|\operatorname{Inv}(f)| = 16$. If we write f in disjoint cycle form, we get $f = (17425)(386)$ so the index $I(f) = 4 + 2 = 6$. Thus, the parity of $I(f)$ (the index) and $|\operatorname{Inv}(f)|$

(the inversion number) are the same in this case. It turns out that they are always the same parity.

To understand why $I(f) \equiv |\mathrm{Inv}(f)| \pmod 2$, we take a close look at Figure 1.75. Look at the adjacent inversion, $(1,2)$, corresponding $f(1) = 7 > f(2) = 5$. Imagine what will happen when you switch 7 and 5 in the one line notation. This switch corresponds to multiplying $f\tau_1 = (17425)(386)(12) = (15)(274)(386)$ or $f\tau_1 = 5\,7\,8\,2\,1\,3\,4\,6$ in one line notation. Note, the dot corresponding to $(1,2)$ is removed from $\mathrm{Inv}(f)$, the others shift around as explained in Figure 1.75 . The new function $f\tau_1$ has an adjacent inversion, $(3,4)$, corresponding to 8 and 2 in one line notation. Remove it by multiplying by an adjacent transposition, $\tau_2 = (34)$ in this case. In this way, you remove one adjacent inversion (from the original set $\mathrm{Inv}(f)$) by one transposition multiplication each time until you have $f\tau_1\tau_2\cdots\tau_q = e$ where $q = |\mathrm{Inv}(f)|$. But we know from Lemma 1.70 that $q \equiv I(f) \pmod 2$. Thus, $I(f) \equiv |\mathrm{Inv}(f)| \pmod 2$. To summarize,

$$(-1)^{I(f)} = (-1)^{|\mathrm{Inv}(f)|} = (-1)^q = \mathrm{sgn}(f). \tag{1.76}$$

Exercises

EXERCISE 1.77. Which permutation on $\underline{n}$ has the most inversions? How many?

EXERCISE 1.78. Show that $\mathrm{sgn}(fg) = \mathrm{sgn}(f)\mathrm{sgn}(g)$ where f and g are permutations on a finite set D. Hint: Use the identity $\mathrm{sgn}(f) = (-1)^q$ where q is the number of transpositions in any representation of f as a product of transpositions . See equation 1.76.

EXERCISE 1.79. In our discussion of Figure 1.75, we proved that the permutation shown there was the product of q adjacent transpositions, $q = |\mathrm{Inv}(f)|$. We assumed that each time we eliminated an inversion, if another inversion remained, we could choose an *adjacent* inversion to be eliminated. Prove that this choice is always possible.

CHAPTER 2

Matrices and Vector Spaces

Review

You should be familiar with Section 1. We summarize a few key definitions: If K and J are sets, then we use $f : K \longrightarrow J$ to indicate that f is a function with domain K and range J. The notation $\{f \mid f : K \longrightarrow J\}$ stands for the set of all functions with domain K and range J (also stated, "set of all f from K to J"). Alternatively, we use exponential notation, J^K, to denote the set of all functions from K to J. For a finite set S, we use $|S|$ to denote the cardinality (number of elements) of S. If S and K are sets then $S \times K = \{(s,t) \mid s \in S, t \in K\}$. We note that $|J^K| = |J|^{|K|}$ if K and J are finite.

REMARK 2.1 (**UNDERLINE NOTATION**). We use the notation $\underline{n} = \{i \mid 1 \leq i \leq n\}$. Thus, $\underline{n} \times \underline{m} = \{(i,j) \mid 1 \leq i \leq n, 1 \leq j \leq m\}$.

DEFINITION 2.2 (**MATRIX**). Let m, n be positive integers. An m by n matrix with entries in a set S is a function $f : \underline{m} \times \underline{n} \longrightarrow S$. The set of all such f is denoted by $\mathbf{M}_{m,n}(S)$. The degenerate case $\mathbf{M}_{0,0}(S) = \emptyset$ is sometimes useful.

In some applications, a matrix is defined as a function $f : \underline{\mu} \times \underline{\nu} \longrightarrow S$ where $\underline{\mu}$ and $\underline{\nu}$ are linearly ordered sets. We won't need that generality in what follows. The subject of "matrix theory" involves the use of matrices and operations defined on them as *conceptual data structures* for understanding various ideas in algebra and combinatorics. For such purposes, matrices have a long history in mathematics and are surprisingly useful.

Next we recall and specialize Definitions 1.24 and 1.27:

DEFINITION 2.3 (**Summary of vector space and algebra axioms**). Let $\mathbb{F}$ be a field and let $(M, +)$ be an abelian group. Assume there is an operation, $\mathbb{F} \times M \longrightarrow M$, which takes (r, x) to rx (juxtaposition of r and x). To show that $(M, +)$ is a *vector space over* $\mathbb{F}$, we show the following four things hold for every $r, s \in \mathbb{F}$ and $x, y \in M$:

(1) $r(x + y) = rx + ry$ (2) $(r + s)x = rx + sx$ (3) $(rs)x = r(sx)$ (4) $1_{\mathbb{F}}\, x = x$

where $1_{\mathbb{F}}$ is the multiplicative identity in $\mathbb{F}$. If $(M, +, \cdot)$ is a ring for which $(M, +)$ is a vector space over $\mathbb{F}$, then $(M, +, \cdot)$ is an *algebra over* $\mathbb{F}$ if the following scalar rule holds:

(5) **scalar rule** for all $\alpha \in \mathbb{F}$, $x, y \in M$ we have $\alpha(xy) = (\alpha x)y = x(\alpha y)$.

The standard example of a vector space is as follows:

Let $J = \mathbb{R}$, the field of real numbers. If f and g are in $\mathbb{R}^K$, we define $h = f + g$ by $h(x) = f(x) + g(x)$ for all $x \in K$. For $\alpha \in \mathbb{R}$ and $f \in \mathbb{R}^K$, we define the product, $h = \alpha f$, of the number α and the function f by $h(x) = \alpha f(x)$ for all $x \in K$. Alternatively stated, $(\alpha f)(x) = \alpha f(x)$ for x in K ($\alpha\beta$ denotes α times β in $\mathbb{R}$).

The set $M = \mathbb{R}^K$, together with these standard rules for adding and multiplying various things, satisfies the following four conditions (see 1.24) that make M a module over $\mathbb{R}$: For all $r, s \in \mathbb{R}$ and $f, g \in M$:

(1) $r(f + g) = rf + rg$ (2) $(r + s)f = rf + sf$ (3) $(rs)f = r(sf)$ (4) $1f = f$.

Since $\mathbb{R}$ is a field, M is a *vector space* by Definition 1.27.

Next, define the *pointwise* product of $f, g \in \mathbb{R}^K$ by $(fg)(x) = f(x)g(x)$. To show that the vector space $M = \mathbb{R}^K$ with this rule of multiplicaton is an algebra, we need to verify the scalar rule of Definition 1.27:

scalar rule for all $\alpha \in \mathbb{R}$, $f, g \in \mathbb{R}^K$ we have $\alpha(fg) = (\alpha f)g = f(\alpha g)$.

This scalar rule follows trivially from the rules for multiplying functions and scalars. Thus, $\mathbb{R}^K$, under the standard rules for function products and multiplication by scalars, *is an algebra over* $\mathbb{R}$, the field of real numbers.

A vector space $(M, +)$ over $\mathbb{F}$ is a *real* vector space if $\mathbb{F} = \mathbb{R}$ and a complex vector space if $\mathbb{F} = \mathbb{C}$. Examples are $\mathbb{R}^K$ and $\mathbb{C}^K$. As noted above, these vector spaces are algebras under pointwise multiplication of functions.

EXAMPLE 2.4. **Examples** Let V be the vector space of column vectors ($n \times 1$ matrices) with real entries:

$$\begin{bmatrix} a_1 \\ a_2 \\ \vdots \\ a_n \end{bmatrix} + \begin{bmatrix} b_1 \\ b_2 \\ \vdots \\ b_n \end{bmatrix} = \begin{bmatrix} a_1 + b_1 \\ a_2 + b_2 \\ \vdots \\ a_n + b_n \end{bmatrix} \quad \text{and} \quad \mu \begin{bmatrix} a_1 \\ a_2 \\ \vdots \\ a_n \end{bmatrix} = \begin{bmatrix} \mu a_1 \\ \mu a_2 \\ \vdots \\ \mu a_n \end{bmatrix}$$

Or let V be the vector space, $\mathbf{M}_{m,n}(\mathbb{R})$, of $m \times n$ matrices over $\mathbb{R}$ where

$$\begin{bmatrix} a_{1,1} & a_{1,2} & \cdots & a_{1,n} \\ a_{2,1} & a_{2,2} & \cdots & a_{2,n} \\ \vdots & \vdots & \vdots & \vdots \\ a_{m,1} & a_{m,2} & \cdots & a_{m,n} \end{bmatrix} + \begin{bmatrix} b_{1,1} & b_{1,2} & \cdots & b_{1,n} \\ b_{2,1} & b_{2,2} & \cdots & b_{2,n} \\ \vdots & \vdots & \vdots & \vdots \\ b_{m,1} & b_{m,2} & \cdots & b_{m,n} \end{bmatrix} =$$

$$\begin{bmatrix} a_{1,1}+b_{1,1} & a_{1,2}+b_{1,2} & \dots & a_{1,n}+b_{1,n} \\ a_{2,1}+b_{2,1} & a_{2,2}+b_{2,2} & \dots & a_{2,n}+b_{2,n} \\ \vdots & \vdots & \vdots & \vdots \\ a_{m,1}+b_{m,1} & a_{m,2}+b_{m,2} & \dots & a_{m,n}+b_{m,m} \end{bmatrix}$$

and

$$\mu \begin{bmatrix} a_{1,1} & a_{1,2} & \dots & a_{1,n} \\ a_{2,1} & a_{2,2} & \dots & a_{2,n} \\ \vdots & \vdots & \vdots & \vdots \\ a_{m,1} & a_{m,2} & \dots & a_{m,n} \end{bmatrix} = \begin{bmatrix} \mu a_{1,1} & \mu a_{1,2} & \dots & \mu a_{1,n} \\ \mu a_{2,1} & \mu a_{2,2} & \dots & \mu a_{2,n} \\ \vdots & \vdots & \vdots & \vdots \\ \mu a_{m,1} & \mu a_{m,2} & \dots & \mu a_{m,n} \end{bmatrix}.$$

Note that the two examples in 2.4 are of the standard form $V = \mathbb{R}^K$. In the first case, $K = \underline{n} = \{1, 2, \cdots, n\}$ and $f(t) = a_t$ for $t \in K$. In the second example, $K = \underline{n} \times \underline{m}$, a cartesian product of two sets, and $f(i, j) = a_{i,j}$ (usually, $a_{i,j}$ is written a_{ij} without the comma).

Review 2.5 (**Linear algebra concepts**). We review some concepts from a standard first course in linear algebra. Let V be a vector space over $\mathbb{R}$ and let S be a subset of V. The *span* of S, denoted Span(S) or Sp(S), is the set of all finite linear combinations of elements in S. That is, Sp(S) is the set of all vectors of the form $\sum_{x\in S} c_x x$ where $c_x \in \mathbb{R}$ and $|\{x \mid c_x \neq 0\}| < \infty$ (*finite support* condition). Note that Sp(S) is a subspace of V. If Sp(S) $= V$ then S is a *spanning set* for V. S is *linearly independent* over $\mathbb{R}$ if $|\{x|c_x \neq 0\}| < \infty$ and $\sum_{x\in S} c_x x = \theta$ (the zero vector) then $c_x = 0$ for all x. If S is linearly independent and spanning, then S is a *basis* for V. If S is a basis and finite ($|S| < \infty$), then V is *finite dimensional.* The cardinality, $|S|$, of this basis is called the *dimension* of V (any two bases for V have the same cardinality). The zero vector is never a member of a basis. The Span$(x_1, x_2, \ldots, x_n)$ of a *sequence* of vectors is the set $\{\sum_{i=1}^{n} c_i x_i \mid c_i \in \mathbb{R}\}$. A *sequence* of vectors, $(x_1, x_2, \ldots, x_n)$, is linearly independent if

$$c_1x_1 + c_2x_2 + \cdots + c_nx_n = 0$$

implies that $c_1 = c_2 = \cdots = c_n = 0$. Otherwise, the sequence is linearly dependent. Thus, the sequence of nonzero vectors (x, x) is linearly dependent due to the repeated vector x. The set $\{x, x\} = \{x\}$ is linearly independent. However, Span(x, x) = Span$(\{x\})$.

Most vector spaces we study will arise as subspaces of finite dimensional vector spaces already familiar to us. The vector space $V = \mathbb{R}^n$ of n-tuples of real numbers will be a frequently used example (2.3). The elements of V are usually written as $(\alpha_1, \alpha_2, \ldots, \alpha_n)$ where each α_i is a real number. This sequence of real numbers can also be regarded as a function f from the set of indices $\underline{n} = \{1, 2, \ldots, n\}$ to $\mathbb{R}$ where $f(i) = \alpha_i$. In this sense, $V = \mathbb{R}^{\underline{n}}$

(using exponential notation for the set of functions). V can also be denoted by $\times^n\mathbb{R}$.

REMARK 2.6 (**DELTA NOTATION**). The n vectors $e_i = (0, \ldots, 0, 1, 0, \ldots, 0)$, where the single 1 occurs in position i, are a basis (the "standard basis") for $V = \mathbb{R}^n$. In the function notation, $\mathbb{R}^{\underline{n}}$, we can simply say that $e_i(j) = \delta(i = j)$ where δ(**Statement**) is 1 if **Statement** is true and 0 if **Statement** is false (1.1). In this case, **Statement** is "$i = j$". One also sees δ_{ij} for $e_i(j)$.

DEFINITION 2.7 (**Subspace**). Let V be a vector space over $\mathbb{R}$, the real numbers, and H a nonempty subset of V. If for any $x, y \in H$ and $\alpha \in \mathbb{R}$, $x + y \in H$ and $\alpha x \in H$, then H is a *subspace* of V. Similarly, we define a subspace of a vector space V over the complex numbers or any field $\mathbb{F}$. Note that a subspace H of V satisfies (or "inherits") all of the conditions of Definition 2.3 and thus H is itself a vector space over $\mathbb{R}$.

Exercises: subspaces

EXERCISE 2.8. Let K_0 be a subset of K. Let U be all f in $\mathbb{R}^K$, the real valued functions with domain K, such that $f(t) = 0$ for $t \in K_0$. Let W be all f in $\mathbb{R}^K$ such that for $i, j \in K_0$, $f(i) = f(j)$. Show that U and W are subspaces of the vector space $V = \mathbb{R}^K$.

EXERCISE 2.9. In each of the following problems determine whether or not the indicated subset is a subspace of the given vector space. The vector space of n-tuples of real numbers is denoted by $\mathbb{R}^n$ or $\times^n\mathbb{R}$ (Cartesian product or n-tuples).

(1) $V = \mathbb{R}^2$; $H = \{(x, y) \mid y \geq 0\}$ [no]

(2) $V = \mathbb{R}^3$; $H = \{(x, y, z) \mid z = 0\}$ [yes]

(3) $V = \mathbb{R}^2$; $H = \{(x, y) \mid x = y\}$ [yes]

(4) $V = C(K)$ where $C(K)$ is all continuous real valued functions on the interval $K = \{x \mid 0 < x < 2\}$; $H = \{f \mid f \in V, f(1) \in \mathbb{Q}\}$ where $\mathbb{Q}$ is the rational numbers. [no]

(5) $V = C(K)$; H is all $f \in C(K)$ such that $3\frac{df}{dx} = 2f$. [yes]

(6) $V = C(K)$; H is all $f \in C(K)$ such that there exists real numbers α and β (depending on f) such that $\alpha\frac{df}{dx} = \beta f$. [find counter example]

(7) $V = \mathbb{R}^{N_0}$ where $N_0 = \{0, 1, 2, 3, \ldots\}$ is the set of nonnegative integers; $H = \{f \mid f \in V,\ |\{t \mid f(t) \neq 0\}| < \infty\}$ is the set of all functions in V with "finite support." [yes]

EXERCISE 2.10. Let $\mathbf{M}_{n,n}(\mathbb{R})$ be the real vector space of $n \times n$ matrices (Example 2.4). Let H_1 be the subspace of *symmetric* matrices $B = (b_{ij}), 1 \leq i, j \leq n$, where $b_{ij} = b_{ji}$ for all i and j. Let H_2 be the subspace of *lower triangular* matrices $B = (b_{ij}), 1 \leq i, j \leq n$, where $b_{ij} = 0$ for all $i < j$. Show that for any two subspaces, H_1 and H_2, of a vector space V, $H_1 \cap H_2$ is a subspace. What is $H_1 \cap H_2$ for this example? What is the smallest subspace containing $H_1 \cup H_2$ in this example?

Exercises: spanning sets and dimension

EXERCISE 2.11. Show that the matrices

$$\begin{pmatrix} 1 & 1 \\ 0 & 1 \end{pmatrix}, \begin{pmatrix} -1 & 1 \\ 0 & 1 \end{pmatrix}, \begin{pmatrix} 0 & 1 \\ 0 & 0 \end{pmatrix}$$

do not span the vector space of all 2×2 matrices over $\mathbb{R}$.

EXERCISE 2.12. Let V be a vector space and $x_1, x_2, \ldots, x_n$ be a sequence (ordered list) of vectors in V. If

$$x_1 \neq 0, x_2 \notin \text{Sp}(\{x_1\}), x_3 \notin \text{Sp}(\{x_1, x_2\}), \ldots, x_n \notin \text{Sp}(\{x_1, x_2, \ldots x_{n-1}\}),$$

show that the vectors $x_1, x_2, \ldots, x_n$ are linearly independent. [use induction]

EXERCISE 2.13. Let V be a vector space and $x_1, x_2, \ldots, x_n$ be linearly independent vectors in V. Let

$$y = \alpha_1 x_1 + \alpha_2 x_2 + \cdots + \alpha_n x_n .$$

What condition on the scalars α_i will guarantee that for each i, the vectors $x_1, x_2, \ldots, x_{i-1}, y, x_{i+1}, \ldots, x_n$ are linearly independent? [all $\alpha_i \neq 0$]

EXERCISE 2.14. Show that the vectors $s_t = \sum_1^t e_i, t = 1, \ldots, n$, are a basis for $\mathbb{R}^{\underline{n}}$ where $e_i(j) = \delta(i = j), i = 1, \ldots, n$ (see Remark 2.6 for notation).

EXERCISE 2.15. Recall the basics of matrix multiplication. The matrices

$$\sigma_x = \begin{pmatrix} 0 & 1 \\ 1 & 0 \end{pmatrix}, \sigma_y = \begin{pmatrix} 0 & -i \\ i & 0 \end{pmatrix}, \sigma_z = \begin{pmatrix} 1 & 0 \\ 0 & -1 \end{pmatrix}$$

are called the *Pauli spin matrices*. Show that these three matrices plus the identity matrix I_2 (2.39) form a basis for the vector space of 2×2 matrices over the complex numbers, $\mathbb{C}$. Show also that

$$\sigma_x \sigma_y = -\sigma_y \sigma_x \quad \text{and} \quad \sigma_x \sigma_z = -\sigma_z \sigma_x \quad \text{and} \quad \sigma_y \sigma_z = -\sigma_z \sigma_y .$$

Exercise 2.16. Let $A_1, A_2, \ldots, A_k$ be a sequence of $m \times n$ matrices. Let $X \neq \theta_{n\times 1}$ be an $n \times 1$ matrix ($\theta_{n\times 1}$ is the $n \times 1$ zero matrix). Show that if $A_1X = A_2X = \cdots = A_kX = \theta_{m\times 1}$ then the matrices $A_1, A_2, \ldots, A_k$ do not form a basis for the vector space $\mathbf{M}_{m,n}(\mathbb{R})$ of all $m \times n$ matrices.

Exercise 2.17. Find a basis for the vector space of 3×3 matrices, $\mathbf{M}_{3,3}(\mathbb{R})$, that consists only of matrices A that satisfy $A^2 = A$ (these are called *idempotent* matrices). Hint: For the case $n = 2$, here is such a basis:

$$\begin{pmatrix} 1 & 0 \\ 0 & 0 \end{pmatrix} \begin{pmatrix} 0 & 0 \\ 0 & 1 \end{pmatrix} \begin{pmatrix} 0 & 0 \\ 1 & 1 \end{pmatrix} \begin{pmatrix} 0 & 1 \\ 0 & 1 \end{pmatrix}.$$

The *standard basis* for $\mathbf{M}_{n,n}(\mathbb{R})$ is the set of n^2 distinct matrices each with one entry equal to one and the rest zero. It is easy to get the standard basis for $\mathbf{M}_{2,2}(\mathbb{R})$ from the four matrices above. Now consider the four idempotent matrices in $\mathbf{M}_{3,3}(\mathbb{R})$.

$$\begin{pmatrix} 1 & 0 & 0 \\ 0 & 0 & 0 \\ 0 & 0 & 0 \end{pmatrix} \begin{pmatrix} 0 & 0 & 0 \\ 0 & 1 & 0 \\ 0 & 0 & 0 \end{pmatrix} \begin{pmatrix} 0 & 0 & 0 \\ 1 & 1 & 0 \\ 0 & 0 & 0 \end{pmatrix} \begin{pmatrix} 0 & 1 & 0 \\ 0 & 1 & 0 \\ 0 & 0 & 0 \end{pmatrix}.$$

From these you can get four of the nine standard basis elements in $\mathbf{M}_{3,3}(\mathbb{R})$.

Exercise 2.18. Show that if $A_iA_j = A_jA_i$ for all $1 \leq i < j \leq k$, then the $n \times n$ matrices $A_1, A_2, \ldots, A_k$ are not a basis for $\mathbf{M}_{n,n}(\mathbb{R})$ ($n > 1$).

Exercise 2.19. (**Trace of matrix**) Show that if $A_i = B_iC_i - C_iB_i$ for $1 \leq i \leq k$, where A_i, B_i and C_i are $n \times n$ matrices, then the $A_1, A_2, \ldots, A_k$ do not form a basis for the vector space $\mathbf{M}_{n,n}(\mathbb{R})$ ($n > 1$). Hint: Recall that the trace of a matrix A, $\text{Tr}(A)$, is the sum of the diagonal entries of A (i.e., $A(1,1)+A(2,2)+\cdots+A(n,n)$). It is easy to show that $\text{Tr}(A + B) = \text{Tr}(A) + \text{Tr}(B)$ and $\text{Tr}(AB) = \text{Tr}(BA)$. This latter fact, $\text{Tr}(AB) = \text{Tr}(BA)$, implies that if A and B are similar matrices, $A = SBS^{-1}$, then $\text{Tr}(A) = \text{Tr}(B)$ (a fact not needed for this exercise). The proof is trivial: $\text{Tr}(S(BS^{-1})) = \text{Tr}((BS^{-1})S) = \text{Tr}(B)$.

Exercise 2.20. Is it possible to span the vector space $\mathbf{M}_{n,n}(\mathbb{R})$ ($n > 1$) using the powers of a single matrix: $I_n, A, A^2, \ldots, A^t, \ldots$?

Exercise 2.21. Show that any $n \times n$ matrix with real coefficients satisfies a polynomial equation $f(A) = 0$, where $f(x)$ is a nonzero polynomial with real coefficients. Hint: Can the matrices $I_n, A, A^2, \ldots, A^p$ be linearly independent for all p?

Matrices – basic stuff

We first discuss some notational issues regarding matrices.

Remark 2.22 (**Index-to-entry function**). The matrix A in 2.23 is shown in the standard general form for an $m \times n$ matrix. Assume the entries of A are from some set S. From 2.2, A is the function $(i, j) \mapsto a_{ij}$ whose domain is $\underline{m} \times \underline{n}$ and whose range is S. As we shall see, matrices can be interpreted as functions in other ways. We refer to the basic definition (2.2) as the index-to-entry representation of A. The standard rectangular presentation is as follows:

$$A = \begin{bmatrix} a_{11} & a_{12} & \dots & a_{1n} \\ a_{21} & a_{22} & \dots & a_{2n} \\ \vdots & \vdots & \vdots & \vdots \\ a_{m1} & a_{m2} & \dots & a_{mn} \end{bmatrix} \tag{2.23}$$

In 2.24 we see two representations of the same index-to-entry function, A. The first representation is the standard two-line description of a function. The domain, $\underline{2} \times \underline{2}$, is listed in lexicographic order as the first line; the values of the function are the second line. In the second representation, the domain values are not shown explicitly but are inferred by the standard rule for indexing the elements of a matrix, $(i, j) \longrightarrow A(i, j)$, where i is the row index and j the column index. Thus, $A(1, 1) = 2$, $A(1, 2) = 2$, $A(2, 1) = 3$, $A(2, 2) = 4$.

$$A = \begin{pmatrix} (1,1) & (1,2) & (2,1) & (2,2) \\ 2 & 2 & 3 & 4 \end{pmatrix} \quad \text{or} \quad A = \begin{pmatrix} 2 & 2 \\ 3 & 4 \end{pmatrix} \tag{2.24}$$

The second representation of the *index-to-entry* function A in 2.24 is the one most used in matrix theory.

Remark 2.25 (**Matrices and function composition**). Matrices as index-to-entry functions can be composed (1.44) with other functions. Here is A, first in two line and then in standard matrix form:

$$A = \begin{pmatrix} (1,1) & (1,2) & (2,1) & (2,2) \\ 2 & 2 & 3 & 4 \end{pmatrix} = \begin{pmatrix} 2 & 2 \\ 3 & 4 \end{pmatrix}, \tag{2.26}$$

Next is a permutation σ of $\underline{2} \times \underline{2}$:

$$\sigma = \begin{pmatrix} (1,1) & (1,2) & (2,1) & (2,2) \\ (1,2) & (2,1) & (2,2) & (1,1) \end{pmatrix}. \tag{2.27}$$

Next compose A with σ (work with two line forms):

$$A \circ \sigma \equiv A\sigma = \begin{pmatrix} (1,1) & (1,2) & (2,1) & (2,2) \\ 2 & 3 & 4 & 2 \end{pmatrix} = \begin{pmatrix} 2 & 3 \\ 4 & 2 \end{pmatrix}. \tag{2.28}$$

Compose σ with a second permutation τ

$$\tau = \begin{pmatrix} (1,1) & (1,2) & (2,1) & (2,2) \\ (1,1) & (2,1) & (1,2) & (2,2) \end{pmatrix} \tag{2.29}$$

to obtain $\sigma\tau$ in two line notation:

$$\sigma\tau = \begin{pmatrix} (1,1) & (1,2) & (2,1) & (2,2) \\ (1,2) & (2,2) & (2,1) & (1,1) \end{pmatrix}. \tag{2.30}$$

Finally, compose A with $\sigma\tau$:

$$A(\sigma\tau) = (A\sigma)\tau = \begin{pmatrix} (1,1) & (1,2) & (2,1) & (2,2) \\ 2 & 4 & 3 & 2 \end{pmatrix} = \begin{pmatrix} 2 & 4 \\ 3 & 2 \end{pmatrix}. \tag{2.31}$$

Converting σ (2.27) to cycle notation, we get $\sigma = ((1,1),(1,2),(2,1),(2,2))$ which is a cycle of length four. In cycle notation, τ (2.29) is $\tau = ((1,2),(2,1))$ which is a transposition. In fact, $A\tau$ is called the *transpose* (3.6) of A. The permutation $\sigma\tau = ((1,1),(1,2),(2,2))$ is a three cycle.
Note that the matrix A (2.24) can be composed with functions that are not permutations:

$$f = \begin{pmatrix} (1,1) & (1,2) \\ (1,2) & (2,1) \end{pmatrix} \quad Af = \begin{pmatrix} (1,1) & (1,2) \\ 2 & 3 \end{pmatrix} = \begin{pmatrix} 2 & 3 \end{pmatrix}. \tag{2.32}$$

In 2.32, the function f transforms a 2×2 matrix A into a 1×2 matrix Af.

REMARK 2.33 (**FUNCTION TERMINOLOGY APPLIED TO MATRICES**). Consider the matrix A of 2.25:

$$A = \begin{pmatrix} (1,1) & (1,2) & (2,1) & (2,2) \\ 2 & 2 & 3 & 4 \end{pmatrix} = \begin{pmatrix} 2 & 2 \\ 3 & 4 \end{pmatrix} \tag{2.34}$$

Recall the terminology for functions associated with Definition 1.38 through Definition 1.47 (coimage). The domain(A) $= \{(1,1),(1,2),(2,1),(2,2)\}$. The image($A$) $= \{2,3,4\}$. The range of A could be any set containing the image (e.g., $\underline{4}$). The coimage(A) $= \{\{(1,1),(1,2)\},\{(2,1)\},\{(2,2)\}\}$.

DEFINITION 2.35 (**BASIC MATRIX NOTATIONAL CONVENTIONS**). Let $A = (a_{ij})$ and $A' = (a'_{ij})$ be two $m \times n$ matrices with entries in a set S. We use the notation $A(i,j) \equiv a_{ij}$. Two matrices are equal, $A = A'$, if $A(i,j) = A'(i,j)$ for all $(i,j) \in \underline{m} \times \underline{n}$. $A_{(i)} = (a_{i1} \dots a_{in})$, $1 \le i \le m$, designates row i of A. $A_{(i)}$ is a $1 \times n$ matrix called a *row vector* of A. $A^{(j)}$, $1 \le j \le n$, designates column j of A. $A^{(j)}$ is an $m \times 1$ matrix called a *column vector* of A.

The range S of the index-to-entry function of a matrix can be quite general. Figure 2.36 shows two matrices, N and C, which have matrices as entries.

(2.36) **Figure : Matrices with matrices as entries :**

$$N = \begin{bmatrix} [a_{11} & a_{12} & \dots & a_{1n}] \\ [a_{21} & a_{22} & \dots & a_{2n}] \\ \vdots & \vdots & \vdots & \vdots \\ [a_{m1} & a_{m2} & \dots & a_{mn}] \end{bmatrix} \qquad C = \begin{bmatrix} \begin{bmatrix} a_{11} \\ a_{21} \\ \vdots \\ a_{m1} \end{bmatrix} & \begin{bmatrix} a_{12} \\ a_{22} \\ \vdots \\ a_{m2} \end{bmatrix} & \begin{matrix} \dots \\ \dots \\ \vdots \\ \dots \end{matrix} & \begin{bmatrix} a_{1n} \\ a_{2n} \\ \vdots \\ a_{mn} \end{bmatrix} \end{bmatrix}$$

The matrix N (2.36) is an $m \times 1$ matrix with each entry a $1 \times n$ row vector of a matrix A (2.35). Thus, $N(i, 1)$, $1 \leq i \leq m$, is the $1 \times n$ *matrix* consisting of row i of A which, in terms of A, is designated $A_{(i)}$.

The matrix C (2.36) is a $1 \times n$ matrix with each entry an $m \times 1$ matrix called a *column vector* of A. Thus, $C(1, j)$, $1 \leq j \leq n$, is the $m \times 1$ *matrix* consisting of column j of A which, in terms of A, is designated $A^{(j)}$.

Remark 2.37 (**Equality of row and column vectors**). If A is an $m \times n$ matrix, a row vector $A_{(i)}$ can never be equal to a column vector $A^{(j)}$ (unless $m = n = 1$). Two matrices can be equal only if they have the same number of rows and the same number of columns. Sometimes you will see a statement that "row i equals column j." Such a statement might be made, for example, if $m = n$ and the sequence of numbers in $A_{(i)}$ equals the sequence of numbers in $A^{(j)}$. In this case the row vector equals the column vector as sequences of numbers, not as matrices.

(2.38) **Multiplication of matrices**

The relationship between linear transformations and matrices is the primary (but not only) motivation for the following definition of matrix multiplication:

Definition 2.39 (**Matrix multiplication**). Let $A \in \mathbf{M}_{m,p}(\mathbb{K})$ and $B \in \mathbf{M}_{p,n}(\mathbb{K})$ be matrices (1.30 for $\mathbb{K}$ of interest here). Let $A(i,j) = a_{ij}$, $(i,j) \in \underline{m} \times \underline{p}$, and $B(i,j) = b_{ij}$, $(i,j) \in \underline{p} \times \underline{n}$. The product $D = AB$ is defined by

(2.40) $$d_{ij} = D(i,j) = \sum_{k=1}^{p} A(i,k)B(k,j) = \sum_{k=1}^{p} a_{ik}b_{kj}, \quad (i,j) \in \underline{m} \times \underline{n}.$$

The $q \times q$ matrix I_q, defined by $I_q(i,j) = 1$ if $i = j$ and $I_q(i,j) = 0$ if $i \neq j$, is called the $q \times q$ *identity matrix*. For an $m \times n$ matrix M,

(2.41) $$I_m M = M I_n = M.$$

From Definition 2.35 we have

(2.42) $$(M_{(i)})^{(j)} = (M^{(j)})_{(i)} \equiv M_{(i)}^{(j)} = (M(i,j)).$$

Another way to write the sum in Definition 2.39 is using the *summation convention*

$$D(i,j) = A(i,k)B(k,j) \tag{2.43}$$

where the two consecutive k indices imply the summation.

It is easy to show that if A is $n_0 \times n_1$, B and C are $n_1 \times n_2$ and D is $n_2 \times n_3$, then the *distributive laws* hold:

$$A(B+C) = AB + AC \text{ and } (B+C)D = BD + CD. \tag{2.44}$$

The associative law for matrix multiplication

An important property of matrix multiplication is that it is *associative*. If A is an $n_0 \times n_1$ matrix, B an $n_1 \times n_2$ matrix, and C an $n_2 \times n_3$ matrix, then $(AB)C = A(BC)$. An aficionado of the summation convention would give a short proof:

$$((AB)C)(i,j) = (A(i,t_1)B(t_1,t_2))C(t_2,j)$$
$$= A(i,t_1)(B(t_1,t_2)C(t_2,j)) = (A(BC))(i,j).$$

Here is the longer proof of the associative law for matrix multiplication using explicit summation notation:

$$((AB)C)(i,j) = \sum_{t_2=1}^{n_2}\left(\sum_{t_1=1}^{n_1} A(i,t_1)B(t_1,t_2)\right)C(t_2,j) = \tag{2.45}$$

$$\sum_{t_2=1}^{n_2}\sum_{t_1=1}^{n_1} A(i,t_1)B(t_1,t_2)C(t_2,j) =$$

$$\sum_{t_1=1}^{n_1}\sum_{t_2=1}^{n_2} A(i,t_1)B(t_1,t_2)C(t_2,j) =$$

$$\sum_{t_1=1}^{n_1} A(i,t_1)\left(\sum_{t_2=1}^{n_2} B(t_1,t_2)C(t_2,j)\right) = (A(BC))(i,j).$$

The associativity of matrix multiplication is a powerful combinatorial tool. Suppose we are to compute the product $C = A_1A_2A_3A_4$ of four matrices assuming, of course, that the product is defined. For example, suppose A_1 is $n \times n_1$, A_2 is $n_1 \times n_2$, A_3 is $n_2 \times n_3$, A_4 is $n_3 \times m$. We can express $C(i,j)$, an entry in the $n \times m$ matrix C, as

$$C(i,j) = \sum_{t_1,t_2,t_3} A_1(i,t_1)A_2(t_1,t_2)A_3(t_2,t_3)A_4(t_3,j)$$

where the sum is over all $(t_1,t_2,t_3) \in \underline{n}_1 \times \underline{n}_2 \times \underline{n}_3$ in any order.

Let A, L and R be $n \times n$ matrices. Suppose that $LA = I_n$ and $AR = I_n$. Then $R = (LA)R = L(AR) = L$, and, hence, $R = L$ (see 1.5). The matrix $B = R = L$ is called the *inverse* of A if it exists. We use the notation, $B = A^{-1}$ for the inverse of A:

(2.46) $$A^{-1}A = AA^{-1} = I_n.$$

If a matrix A has an inverse, we say that A is *nonsingular* or *invertible* or a *unit* in the ring $R = \mathbf{M}_{n,n}(\mathbb{K})$ (1.9).

Again, assume $D = AB$ where A is an $m \times p$ matrix and B a $p \times n$ matrix. Note that Definition 2.39 also implies that the $1 \times n$ row matrix $D_{(i)}$ and the $m \times 1$ column matrix $D^{(j)}$ satisfy

(2.47) $$D_{(i)} = (AB)_{(i)} = A_{(i)}B \quad \text{and} \quad D^{(j)} = (AB)^{(j)} = AB^{(j)}.$$

Explicitly, for the row version we have

(2.48) $$D_{(i)} = A_{(i)}B = \sum_{t=1}^{p} A(i,t)B_{(t)}.$$

The column version of 2.48 is

(2.49) $$D^{(j)} = AB^{(j)} = \sum_{t=1}^{p} A^{(t)}B(t,j).$$

Equation 2.48 states that row i of the product AB is a linear combination of the rows of B with coefficients from row i of A. Equation 2.49 states that column j of the product AB is a linear combination of the columns of A with coefficients from column j of B. Here is an example:

(2.50) **Figure : Matrix times a vector**

$$A = \begin{pmatrix} 2 & 1 & 3 \\ 1 & -1 & 0 \\ 1 & 0 & 2 \end{pmatrix} \qquad X = \begin{pmatrix} 1 \\ 2 \\ 3 \end{pmatrix} \qquad Y = (1 \ 2 \ 3)$$

$$AX = 1\begin{pmatrix} 2 \\ 1 \\ 1 \end{pmatrix} + 2\begin{pmatrix} 1 \\ -1 \\ 0 \end{pmatrix} + 3\begin{pmatrix} 3 \\ 0 \\ 2 \end{pmatrix} = \begin{pmatrix} 13 \\ -1 \\ 7 \end{pmatrix}$$

The result is a 3 x 1 column vector

$$\mathrm{YA} = \left.\begin{array}{l} 1\,(2 \quad 1 \quad 3) = (2 \ \ 1 \ \ 3) \\ 2\,(1 \ \ {-1} \quad 0) = (2 \ \ {-2} \ \ 0) \\ 3\,(1 \quad 0 \quad 2) = (3 \ \ 0 \ \ 6) \end{array}\right\}$$

Add to get (7 -1 9)

We need some notation for *submatrices* of a matrix. Figure 2.51 gives some examples ($p = 5$, $n = 3$) of what is needed. Note that f and g are functions with domain $\underline{3}$ and range $\underline{5}$ (i.e., elements of $\underline{5}^{\underline{3}}$). The function c is a permutation of $\underline{3}$ and fc denotes the composition of f and c (i.e., $fc \in \underline{5}^{\underline{3}}$).

(2.51) **Figure : Submatrix notation : $f, g \in \underline{5}^{\underline{3}}$ and $c = (1\,3)$.**

$$g = \begin{pmatrix} 1 & 2 & 3 \\ 5 & 1 & 4 \end{pmatrix} \qquad f = \begin{pmatrix} 1 & 2 & 3 \\ 1 & 3 & 5 \end{pmatrix} \quad c = (1\,3) \quad fc = \begin{pmatrix} 1 & 2 & 3 \\ 5 & 3 & 1 \end{pmatrix}$$

$$A = \begin{bmatrix} 2 & 1 & 3 & 1 & 1 \\ 0 & 3 & 1 & 1 & 2 \\ 1 & 1 & 1 & 2 & 1 \end{bmatrix} \quad \overset{A^{(5)}A^{(1)}A^{(4)}}{A^g = \begin{bmatrix} 1 & 2 & 1 \\ 2 & 0 & 1 \\ 1 & 1 & 2 \end{bmatrix}} \quad A^f = \begin{bmatrix} 2 & 3 & 1 \\ 0 & 1 & 2 \\ 1 & 1 & 1 \end{bmatrix} \quad A^{fc} = \begin{bmatrix} 1 & 3 & 2 \\ 2 & 1 & 0 \\ 1 & 1 & 1 \end{bmatrix}$$

What we call "submatrices" is an extension of the usual usage. Here is the formal definition:

Definition 2.52 (**Submatrix notation**). Let X be an $m \times n$ matrix and let $f \in \underline{m}^{\underline{r}}$ and $g \in \underline{n}^{\underline{s}}$ be functions, $r, s \geq 0$. We define the $r \times s$ matrix

(2.53) $$Y \equiv X_f^g \equiv X[f \mid g] \equiv X[f(1), \ldots, f(r) \mid g(1), \ldots, g(s)]$$

by $Y(p,q) = X(f(p), g(q))$. If $r = m$ and f the identity we write X_f^g as X^g. Suppose $\alpha = \{a_1, \ldots, a_r\} \subseteq \underline{m}$ and $\beta = \{b_1, \ldots, b_s\} \subseteq \underline{n}$ are subsets of size r and s where the $a_1 < \cdots < a_r$ and $b_1 < \cdots < b_s$ are in increasing order. Define

(2.54) $$Y \equiv X[\alpha \mid \beta] \equiv X[a_1, \ldots, a_r \mid b_1, \ldots, b_s]$$

by $Y(i,j) = X(a_i, b_j)$ for all $(i,j) \in \underline{r} \times \underline{s}$. If $\alpha' = \underline{m} - \alpha$ and $\beta' = \underline{n} - \beta$ are the complements of α and β, then define

(2.55) $$X(\alpha \mid \beta) = X[\alpha' \mid \beta'],\ X(\alpha \mid \beta] = X[\alpha' \mid \beta],\ X[\alpha \mid \beta) = X[\alpha \mid \beta'].$$

Note that in Definition 2.52 we have

(2.56) $$(X^g)_f = (X_f)^g = X_f^g.$$

Remark 2.57 (**Example of submatrix notation**). Let $m = 2$ and $n = 3$ with

(2.58) $$X = \begin{pmatrix} 1 & 2 & 3 \\ 4 & 5 & 6 \end{pmatrix}.$$

Let $r = 3$ and $s = 4$ with $f = (1\,2\,1)$ and $g = (2\,3\,2\,1)$. Then

(2.59) $$Y = X[f \mid g] = \begin{pmatrix} 2 & 3 & 2 & 1 \\ 5 & 6 & 5 & 4 \\ 2 & 3 & 2 & 1 \end{pmatrix}.$$

From equation 2.59 we note that $Y = X[f \mid g]$ then $Y(2,4) = 4$ but $X(2,1) = 4$.

Remark 2.60 (**Submatrices as sets or functions**). Note that 2.54 is a special case of 2.53. Let α and β be as in 2.54, and define $f_\alpha \in \mathrm{SNC}(r,m)$ and $f_\beta \in \mathrm{SNC}(s,n)$ (1.50) by image$(f_\alpha) = \alpha$ and image$(f_\beta) = \beta$. Then

$$X[\alpha \mid \beta] = X[a_1, \ldots, a_r \mid b_1, \ldots, b_s] = X[f_\alpha \mid f_\beta].$$

Using Definition 2.52, we can generalize identity 2.47. Assume now that $D = AB$ where A is an $a \times p$ matrix and B a $p \times b$ matrix. Let $g \in \underline{a}^{\underline{m}}$ and $h \in \underline{b}^{\underline{n}}$ be functions. We can think of g as a "row selection" function so that A_g is an $m \times p$ matrix and h as a "column selection" function so that B^h is a $p \times n$ matrix. Thus, the product $A_g B^h$ is a $m \times n$ matrix. Then, we have

$$D[g \mid h] \equiv D_g^h \equiv (AB)_g^h = A_g B^h. \tag{2.61}$$

CHAPTER 3

Determinants

We now define the determinant of an $n \times n$ matrix. In the discussion of determinants that follows, assume the matrices have entries in the rings, $\mathbb{K}$, described in Remark 1.30, all of which are Euclidean domains. If you are interested in more generality, review the discussions of Section 1, specifically 1.8, 1.9, 1.14, 1.16, and do a web search for "rings determinants." Recall the definition of the sign, $\operatorname{sgn}(f)$, of a permutation (Definition 1.72) and the discussion that follows that definition, including identity 1.76

DEFINITION 3.1 (**Determinant**). Let A be an $n \times n$ matrix with entries $A(i, j)$. The determinant, $\det(A)$, is defined by

$$\det(A) = \sum_{f} \operatorname{sgn}(f) \prod_{i=1}^{n} A(i, f(i))$$

where the sum is over all permutations of the set $\underline{n} = \{1, 2, \ldots, n\}$.

The terms of the product, $\prod_{i=1}^{n} A(i, f(i))$, commute. Thus, the product can be taken in any order over the set, $\operatorname{Graph}(f)$ (1.39):

$$\det(A) = \sum_{f} \operatorname{sgn}(f) \prod_{(i,j)\in\operatorname{Graph}(f)} A(i, j). \tag{3.2}$$

In particular, note that

$$\operatorname{Graph}(f) = \{(i, f(i)) \mid i \in \underline{n}\} = \{(f^{-1}(i), i) \mid i \in \underline{n}\}. \tag{3.3}$$

Thus, we have

$$\det(A) = \sum_{f} \operatorname{sgn}(f) \prod_{\operatorname{Graph}(f)} A(i, j) = \sum_{f} \operatorname{sgn}(f) \prod_{i=1}^{n} A(f^{-1}(i), i). \tag{3.4}$$

Summing over all f is the same as summing over all f^{-1} and using $\operatorname{sgn}(f) = \operatorname{sgn}(f^{-1})$ (which follows from $ff^{-1} = e$ and $\operatorname{sgn}(e) = +1$). Therefore, the second sum in 3.4 can be written

$$\sum_{f^{-1}} \operatorname{sgn}(f^{-1}) \prod_{i=1}^{n} A(f^{-1}(i), i) = \sum_{f} \operatorname{sgn}(f) \prod_{i=1}^{n} A(f(i), i).$$

Thus, we have the important identity

$$\det(A) = \sum_f \operatorname{sgn}(f) \prod_{i=1}^{n} A(i, f(i)) = \sum_f \operatorname{sgn}(f) \prod_{i=1}^{n} A(f(i), i). \tag{3.5}$$

The first sum in 3.5 is called the row form of the determinant and the second is called the column form. In the first sum, the domain of f is the set of row indices and the range is the set of column indices. In the second, the domain is the set of column indices and the range is the set of row indices.

DEFINITION 3.6 (**TRANSPOSE OF A MATRIX**). Let A be an $n \times n$ matrix with entries $A(i, j)$. The *transpose* of A is the matrix A^T defined by $A^T(i, j) = A(j, i)$. This definition extends naturally to A an $n \times m$ matrix.

REMARK 3.7 (**TRANSPOSE BASICS**). The transpose of $A = \begin{pmatrix} a_{11} & a_{12} \\ a_{21} & a_{22} \end{pmatrix}$ is $A^T = \begin{pmatrix} a_{11} & a_{21} \\ a_{12} & a_{22} \end{pmatrix}$. Note that the transpose of a product is the product of the transposes in reverse order: $(AB)^T = B^T A^T$. Recall remark 2.22 concerning the index-to-entry function and note that $A^T(2, 1) = a_{12} = A(1, 2)$. Suppose we take $A = (a_{ij})$ to be a 4×4 matrix. Let $X = A(2 \mid 3)$ be the 3×3 submatrix

$$X = \begin{pmatrix} a_{11} & a_{12} & a_{14} \\ a_{31} & a_{32} & a_{34} \\ a_{41} & a_{42} & a_{44} \end{pmatrix} \text{ and } X^T = \begin{pmatrix} a_{11} & a_{31} & a_{41} \\ a_{12} & a_{32} & a_{42} \\ a_{14} & a_{34} & a_{44} \end{pmatrix}.$$

Recall remark 2.22 concerning the index-to-entry function and note that $X(1, 3) = a_{14}$ and $X^T(1, 3) = a_{41}$. The index-to-entry function has domain $\underline{3} \times \underline{3}$ for these submatrices. Thus, $X(i, j) = X^T(j, i)$, $1 \leq i, j \leq 3$ as required by definition 3.6. Starting with A, we have $(A(2 \mid 3))^T = A^T(3 \mid 2)$. You can first take a submatrix of A and then transpose that or first transpose A and then take the appropriate submatrix. Using the submatrix notation of 2.52, the rule is

$$(X[f \mid g])^T = X^T[g \mid f]. \tag{3.8}$$

As an example, consider

$$X = \begin{pmatrix} 1 & 2 & 3 \\ 4 & 5 & 6 \end{pmatrix} \text{ and } X^T = \begin{pmatrix} 1 & 4 \\ 2 & 5 \\ 3 & 6 \end{pmatrix}.$$

Let $r = 3$ and $s = 4$ with $f = (1\,2\,1)$ and $g = (2\,3\,2\,1)$. Then

$$(X[f \mid g])^T = \begin{pmatrix} 2 & 3 & 2 & 1 \\ 5 & 6 & 5 & 4 \\ 2 & 3 & 2 & 1 \end{pmatrix}^T = \begin{pmatrix} 2 & 5 & 2 \\ 3 & 6 & 3 \\ 2 & 5 & 2 \\ 1 & 4 & 1 \end{pmatrix} = X^T[g \mid f]. \tag{3.9}$$

THEOREM 3.10 (**DETERMINANT OF TRANSPOSE**). *Let A be an $n\times n$ matrix with entries $A(i,j)$ and let A^T be its transpose. Then*

$$\det(A) = \det(A^T).$$

PROOF. We use 3.5.

$$\det(A) = \sum_f \operatorname{sgn}(f) \prod_{i=1}^{n} A(i, f(i)) = \sum_f \operatorname{sgn}(f) \prod_{i=1}^{n} A^T(f(i), i) = \det(A^T).$$

□

Elementary properties of determinants

We now derive some "elementary" properties of the determinant – properties that follow directly from the definition (3.1) using routine (but not necessarily short) computations. It is common to use Greek letters for permutations so we switch to that convention.

First, note that if φ and γ are permutations of $\underline{n}$ then

$$(3.11)\qquad \prod_{i=1}^{n} A(i, \varphi(i)) = \prod_{i=1}^{n} A(\gamma(i), \varphi(\gamma(i))) = \prod_{(i,j)\in \mathrm{Graph}(\varphi)} A(i,j).$$

Let A be an $n \times n$ matrix and let γ be a permutation on $\underline{n}$. As a sequence of columns, we write $A = (A^{(1)}, \ldots, A^{(i)}, \ldots, A^{(n)})$. We define

$$A^\gamma = (A^{(\gamma(1))}, \ldots, A^{(\gamma(i))}, \ldots, A^{(\gamma(n))}),$$

This notation is a special case of Definition 2.52 ($r = s = m = n, g = \gamma, f$ the identity). From 3.5,

$$\det(A^\gamma) = \sum_\varphi \operatorname{sgn}(\varphi) \prod_{i=1}^{n} A^\gamma(\varphi(i), i)$$

where

$$\prod_{i=1}^{n} A^\gamma(\varphi(i), i) = \prod_{i=1}^{n} A(\varphi(i), \gamma(i)) = \prod_{i=1}^{n} A(\varphi\gamma^{-1}(i), i).$$

Thus,

$$(3.12)\qquad \det(A^\gamma) = \sum_\varphi \operatorname{sgn}(\varphi) \prod_{i=1}^{n} A(\varphi\gamma^{-1}(i), i).$$

From 3.12, we get the very important symmetry property of the determinant function under permutation of columns (by a similar argument, rows) which states that $\det(A^\gamma) = \operatorname{sgn}(\gamma)\det(A)$ (for rows, $\det(A_\gamma) = \operatorname{sgn}(\gamma)\det(A)$):

$$(3.13)\qquad \det(A^\gamma) = \operatorname{sgn}(\gamma) \sum_\varphi \operatorname{sgn}(\varphi\gamma^{-1}) \prod_{i=1}^{n} A(\varphi\gamma^{-1}(i), i) = \operatorname{sgn}(\gamma)\det(A).$$

The next definition is fundamental to the study of determinants.

Definition 3.14 (**Multilinear function**). Let $V_1, \ldots, V_n$ be vector spaces over a field $\mathbb{F}$ and let $W = \times_1^n V_i = \{(x_1, \ldots, x_n) \mid x_i \in V_i, i = 1, \ldots, n\}$ be the direct (Cartesian) product of these V_i. A function Φ from W to $\mathbb{F}$ is *multilinear* if it is linear separately in each variable: For $c, d \in \mathbb{F}$ and for $t = 1, \ldots, n$,

$$\Phi(x_1, \ldots, (cx_t + dy_t), \ldots, x_n) = c\Phi(x_1, \ldots, x_t, \ldots, x_n) + d\Phi(x_1, \ldots, y_t, \ldots, x_n).$$

If $n = 1$ then Φ is a *linear function* from V_1 to $\mathbb{F}$.

The determinant is a multilinear function.

If A is an $n \times n$ matrix over $\mathbb{F}$, we can regard A as an ordered sequence, $(x_1, \ldots, x_n)$, of vectors in $\mathbb{F}^n$ where either $x_i = A^{(i)}, i = 1, \ldots, n$, are the columns of A or $x_i = A_{(i)}, i = 1, \ldots, n$, are the rows of A. In either case, rows or columns, $\det(A) = \det(x_1, \ldots, x_n)$ is a multilinear function from $W = \times_1^n V_i$ to $\mathbb{F}$.

To verify multilinearity (row version), let $B_{(t)} = (B(t,1), \ldots, B(t,n))$ for a fixed t. Replace row t of A with $B_{(t)}$, to get $\hat{A}$:

$$\hat{A} = (A_{(1)}, \ldots, B_{(t)}, \ldots, A_{(n)}).$$

Replace row t of A with $cA_{(t)} + dB_{(t)}$, to get $\tilde{A}$:

$$\tilde{A} = (A_{(1)}, \ldots, (cA_{(t)} + dB_{(t)}), \ldots, A_{(n)}).$$

Using the definition of the determinant we compute

$$\det(\tilde{A}) = \sum_{\varphi} \operatorname{sgn}(\varphi) A(1, \varphi(1)) \cdots (cA(t, \varphi(t)) + dB(t, \varphi(t))) \cdots A(n, \varphi(n)) =$$

$$c \sum_{\varphi} \operatorname{sgn}(\varphi) A(1, \varphi(1)) \cdots A(t, \varphi(t)) \cdots A(n, \varphi(n)) +$$

$$d \sum_{\varphi} \operatorname{sgn}(\varphi) A(1, \varphi(1)) \cdots B(t, \varphi(t)) \cdots A(n, \varphi(n)).$$

Thus we have

$$\det(\tilde{A}) = c \det A + d \det(\hat{A}) \tag{3.15}$$

which verifies that the determinant is a multilinear function (3.14).

Definition 3.16 (**Alternating multilinear**). A multilinear function Φ from $\times^n \mathbb{F}$ to $\mathbb{F}$ is *alternating* if for any transposition $\tau = (s\ t)$ on $\underline{n}$,

$$\Phi(x_1, \ldots, x_i, \ldots, x_n) = -\Phi(x_{\tau(1)}, \ldots, x_{\tau(i)}, \ldots, x_{\tau(n)}).$$

Or, equivalently, interchanging x_s and x_t only gives

$$\Phi(x_1, \ldots, x_s, \ldots, x_t, \ldots, x_n) = -\Phi(x_1, \ldots, x_t, \ldots, x_s, \ldots, x_n).$$

In particular, note that if Φ is alternating and $x_s = x_t = x$ then the identities of 3.16 become

$$(3.17) \qquad \Phi(x_1, \ldots, x, \ldots, x, \ldots, x_n) = -\Phi(x_1, \ldots, x, \ldots, x, \ldots, x_n).$$

Thus, $\Phi(x_1, \ldots, x, \ldots, x, \ldots, x_n) = 0$. We use the fact that $\mathbb{F}$ is of characteristic 0 (Definition 1.14 and Remark 1.30). In particular, if any pair of vectors, x_s and x_t, are linearly dependent then

$$(3.18) \qquad \Phi(x_1, \ldots, x_s, \ldots, x_t, \ldots, x_n) = 0.$$

For suppose x_t and x_s are nonzero and $x_t = cx_s$ where $0 \neq c \in \mathbb{F}$. Then

$$\Phi(x_1, \ldots, x_s, \ldots, x_t, \ldots, x_n) = c\Phi(x_1, \ldots, x_s, \ldots, x_s, \ldots, x_n) = 0.$$

The determinant is alternating multilinear

Identity 3.15 shows that $\det(x_1, \ldots, x_n)$ is a multilinear function of its rows or columns. Identity 3.13 implies that if A is an $n \times n$ matrix, γ is a permutation on $\underline{n}$ and (column form) $A^\gamma = (A^{(\gamma(1))}, \ldots, A^{(\gamma(i))}, \ldots, A^{(\gamma(n))})$ or (row form) $A_\gamma = (A_{(\gamma(1))}, \ldots, A_{(\gamma(i))}, \ldots, A_{(\gamma(n))})$ then

$$(3.19) \qquad \det(A^\gamma) = \operatorname{sgn}(\gamma)\det(A) \quad \text{and} \quad \det(A_\gamma) = \operatorname{sgn}(\gamma)\det(A).$$

Thus, if we take $\gamma = \tau = (s\ t)$ we get (using $\operatorname{sgn}(\tau) = -1$)

$$(3.20) \qquad \det(A^{(\tau(1))}, \ldots, A^{(\tau(n))}) = -\det(A^{(1)}, \ldots, A^{(n)}).$$

This identity shows that the determinant is an alternating (multilinear) function of its columns (and, similarly, its rows). Thus, $\det(A) = 0$ if any two rows (or columns) are the same or linearly dependent.

(3.21) General direct sums

We start with the standard definition. See 2.52 for related notation.

Definition 3.22 (**Direct sum of matrices**). We write $\Theta_{p,q}$ for the $p \times q$ zero matrix, p and q nonnegative integers. Let B be an $r \times r'$ matrix and C be $s \times s'$ matrix. The *direct sum* of B and C (written $B \oplus C$) is the unique $(r+s) \times (r'+s')$ matrix A determined by

$$A[\underline{r} \mid \underline{r}'] = B \quad A(\underline{r} \mid \underline{r}') = C \quad A[\underline{r} \mid \underline{r}') = \Theta_{r,s'} \quad A(\underline{r} \mid \underline{r}] = \Theta_{s,r'}.$$

The determinant of a direct sum A of an $r \times r$ matrix B and an $s \times s$ matrix C is the product of the determinants of the summands:

$$(3.23) \qquad \det(B \oplus C) = \det(A) = \det(A[\underline{r} \mid \underline{r}])\det(A(\underline{r} \mid \underline{r})) = \det(B)\det(C).$$

This identity can be proved directly from the definition of the determinant.

As an example of 3.23, let A be a 4×4 matrix of integers ($A \in \mathbf{M}_{4,4}(\mathbb{Z})$) :

$$(3.24) \qquad A = \begin{pmatrix} 1 & 2 & 0 & 0 \\ 2 & 3 & 0 & 0 \\ 0 & 0 & 3 & 4 \\ 0 & 0 & 4 & 1 \end{pmatrix} = \begin{pmatrix} 1 & 2 \\ 2 & 3 \end{pmatrix} \oplus \begin{pmatrix} 3 & 4 \\ 4 & 1 \end{pmatrix}.$$

Then $\det(A) = \det(B)\det(C) = (-1)(-13) = 13$.

Definition 3.25 (**General direct sum**). Let A be an $n \times n$ matrix and let $X, Y \in \mathbb{P}_r(n)$ where $0 \leq r < n$ (size r subsets of $\underline{n}$ 1.32). We say that A is the *general direct sum* relative to X and Y of an $r \times r$ matrix B and $s \times s$ matrix C if $n = r + s$ and

$$A[X \mid Y] = B, \quad A(X \mid Y) = C, \quad A[X \mid Y) = \Theta_{r,s} \text{ and } A(X \mid Y] = \Theta_{s,r}.$$

We write

$$A = B \oplus_X^Y C.$$

As an example of 3.25, let A be a 4×4 matrix of integers ($A \in \mathbf{M}_{4,4}(\mathbb{Z})$) :

$$(3.26) \qquad A = \begin{pmatrix} 1 & 0 & 2 & 0 \\ 0 & 3 & 0 & 4 \\ 2 & 0 & 3 & 0 \\ 0 & 4 & 0 & 1 \end{pmatrix} = \begin{pmatrix} 1 & 2 \\ 2 & 3 \end{pmatrix} \oplus_X^Y \begin{pmatrix} 3 & 4 \\ 4 & 1 \end{pmatrix} \quad X = Y = \{1, 3\}.$$

Note that the direct sum of Definition 3.22 is a special case of Definition 3.25 (take $X = Y = \underline{r}$). The matrix of example 3.26 can be transformed by row and column interchanges to that of 3.24, thus the determinants differ by only the sign. Direct computation gives $\det(A) = 13$ for A in 3.26 . An example of the transformation process by row and column interchanges is given in Figure 3.27. We take a different approach in order to develop precise combinatorial and analytic tools for future use.

(3.27) **Figure : Reduce submatrix $A[X \mid Y]$ to initial position**

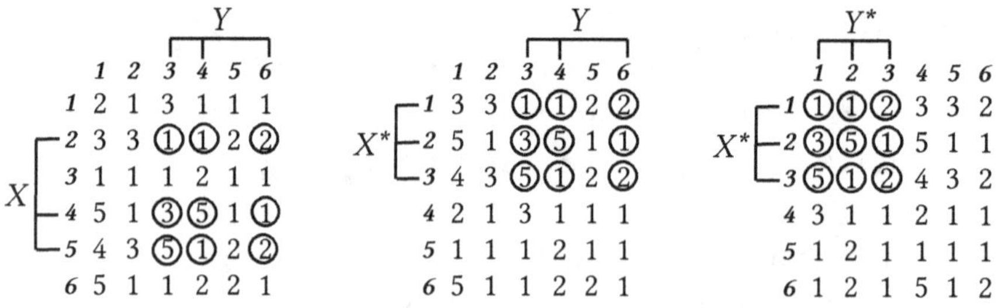

X to X^* in (2-1)+(4-2)+(5-3) =5 row interchanges.

Y to Y^* in (3-1)+(4-2)+(6-3) =7 column interchanges.

Definition 3.28 (**The set S_X^Y**). Let $S = \mathrm{PER}(n)$ be the permutations of $\underline{n}$, and let $X, Y \in \mathbb{P}_k(n)$ be subsets of $\underline{n}$ of size k. Let

$$S_X^Y = \{\sigma \mid \sigma \in S \text{ and } \sigma(X) = Y\}.$$

Let $\gamma \in S_X^Y$ and let X' and Y' denote the complements of X and Y. Suppose the restrictions (1.43) γ_X and $\gamma_{X'}$ of γ are strictly increasing:

$$\gamma_X \in \mathrm{SNC}(X, Y) \quad \text{and} \quad \gamma_{X'} \in \mathrm{SNC}(X', Y').$$

Then γ is called the *canonical* element of S_X^Y. Note that γ_X and $\gamma_{X'}$ are unique since $|X| = |Y|$ and $|X'| = |Y'|$.

Remark 3.29 (**Example of S_X^Y**). We use the notation of 3.28. Take $n = 6$ and $k = 3$, $X, Y \in \mathbb{P}_3(6)$ where $X = \{2, 4, 5\}$ and $Y = \{3, 4, 6\}$. The set S_X^Y consists of all permutations σ such that $\sigma(\{2, 4, 5\}) = \{3, 4, 6\}$ (i.e., $\sigma(X) = Y$). This implies (from the definition of a permutation) that $\sigma(\{1, 3, 6\}) = \{1, 2, 5\}$ (i.e., $\sigma(X') = Y'$). In other words,

$$\mathrm{Graph}(\sigma) \subseteq (X \times Y) \cup (X' \times Y'). \tag{3.30}$$

Remark 3.31 (**Canonical element, γ**). We continue Remark 3.29. Designate the elements of X in order as $(x_1, x_2, x_3) = (2, 4, 5)$ and Y in order as $(y_1, y_2, y_3) = (3, 4, 6)$. Similarly, X' in order is $(x_1', x_2', x_3') = (1, 3, 6)$ and Y' in order is $(y_1', y_2', y_3') = (1, 2, 5)$. In two line notation, let

$$\gamma = \begin{pmatrix} 2 & 4 & 5 & 1 & 3 & 6 \\ 3 & 4 & 6 & 1 & 2 & 5 \end{pmatrix} = \begin{pmatrix} 1 & 2 & 3 & 4 & 5 & 6 \\ 1 & 3 & 2 & 4 & 6 & 5 \end{pmatrix}. \tag{3.32}$$

The permutation γ is the canonical element of S_X^Y (3.28).

Remark 3.33 (**Restrictions**). The γ of 3.32 has the following restrictions to X and X':

$$\gamma_X = \begin{pmatrix} 2 & 4 & 5 \\ 3 & 4 & 6 \end{pmatrix} \text{ and } \gamma_{X'} = \begin{pmatrix} 1 & 3 & 6 \\ 1 & 2 & 5 \end{pmatrix}. \tag{3.34}$$

Note from 3.34 that $\gamma_X \in \mathrm{SNC}(X, Y)$ and $\gamma_{X'} \in \mathrm{SNC}(X', Y')$ as required by the definition of γ in 3.28.

Remark 3.35 (**Typical element**). A typical element $\sigma \in S_X^Y$ has restrictions to X and X' that are bijections:

$$\sigma_X = \begin{pmatrix} 2 & 4 & 5 \\ 6 & 3 & 4 \end{pmatrix} \text{ and } \sigma_{X'} = \begin{pmatrix} 1 & 3 & 6 \\ 5 & 2 & 1 \end{pmatrix}. \tag{3.36}$$

Using the fact that $(y_1, y_2, y_3) = (3, 4, 6)$ and $(y_1', y_2', y_3') = (1, 2, 5)$, the second lines of σ_X and $\sigma_{X'}$ (3.36) can be specified as permutations of (y_1, y_2, y_3) and (y_1', y_2', y_3') :

$$(6, 3, 4) = (y_3, y_1, y_2) = (y_{\nu(1)}, y_{\nu(2)}, y_{\nu(3)}) \quad \nu = \begin{pmatrix} 1 & 2 & 3 \\ 3 & 1 & 2 \end{pmatrix} \tag{3.37}$$

and

$$(5,2,1) = (y_3', y_2', y_1') = (y'_{\mu(1)}, y'_{\mu(2)}, y'_{\mu(3)}) \quad \mu = \begin{pmatrix} 1 & 2 & 3 \\ 3 & 2 & 1 \end{pmatrix}. \tag{3.38}$$

LEMMA 3.39 (**DESCRIPTION OF S_X^Y**). *We use 3.28. Let X be ordered $x_1 < \cdots < x_k$ (i.e., the sequence $x_1, \ldots, x_k$ is ordered as integers) and X' ordered $x_1' < \cdots < x'_{n-k}$. Similarly, let Y be ordered $y_1 < \cdots < y_k$ and Y' ordered $y_1' < \cdots < y'_{n-k}$. Let γ be the canonical element of S_X^Y.*

$$\gamma = \begin{pmatrix} x_1 & \ldots & x_k & x_1' & \ldots & x'_{n-k} \\ y_1 & \ldots & y_k & y_1' & \ldots & y'_{n-k} \end{pmatrix}.$$

For $\sigma \in S_X^Y$, define $\nu \in \mathrm{PER}(k)$ by $(\sigma(x_1), \ldots, \sigma(x_k)) = (y_{\nu(1)}, \ldots, y_{\nu(k)})$. Likewise, define $\mu \in \mathrm{PER}(n-k)$ by $(\sigma(x_1'), \ldots, \sigma(x'_{n-k})) = (y'_{\mu(1)}, \ldots, y'_{\mu(n-k)})$. Then S_X^Y is the set of all permutations of the form

$$\gamma^{\nu\mu} = \begin{pmatrix} x_1 & \ldots & x_k & x_1' & \ldots & x'_{n-k} \\ y_{\nu(1)} & \ldots & y_{\nu(k)} & y'_{\mu(1)} & \ldots & y'_{\mu(n-k)} \end{pmatrix} \tag{3.40}$$

for $\nu \in \mathrm{PER}(k)$ and $\mu \in \mathrm{PER}(n-k)$.

PROOF. Note $\gamma \in S_X^Y$ sends X to Y in some order and likewise X' to Y'. Moreover, $\gamma_X \in \mathrm{SNC}(X, Y)$ and $\gamma_{X'} \in \mathrm{SNC}(X', Y')$ implies these orders must be $y_1, \ldots, y_k$ and $y_1', \ldots, y'_{n-k}$ respectively. This implies the two line notation for γ. For any $\nu \in \mathrm{PER}(k)$ and $\mu \in \mathrm{PER}(n-k)$ the map $\gamma^{\nu\mu} \in S_X^Y$. Conversely, each $\sigma \in S_X^Y$ can be written in the form $\gamma^{\nu\mu}$ for some $\nu \in \mathrm{PER}(k)$ and $\mu \in \mathrm{PER}(n-k)$, as stated in the lemma. Hence, $S_X^Y = \{\gamma^{\nu\mu} \mid \nu \in \mathrm{PER}(k) \text{ and } \mu \in \mathrm{PER}(n-k)\}$.

□

LEMMA 3.41 (**SIGNS OF S_X^Y ELEMENTS**). *We use the terminology of Lemma 3.39. Let X be ordered as integers $x_1, \ldots, x_k$ and X' ordered $x_1', \ldots, x'_{n-k}$. Similarly, let Y be ordered $y_1, \ldots, y_k$ and Y' ordered $y_1', \ldots, y'_{n-k}$. Let γ be the canonical representative.*

$$\gamma = \begin{pmatrix} x_1 & \ldots & x_k & x_1' & \ldots & x'_{n-k} \\ y_1 & \ldots & y_k & y_1' & \ldots & y'_{n-k} \end{pmatrix},$$

and let

$$\gamma^{\nu\mu} = \begin{pmatrix} x_1 & \ldots & x_k & x_1' & \ldots & x'_{n-k} \\ y_{\nu(1)} & \ldots & y_{\nu(k)} & y'_{\mu(1)} & \ldots & y'_{\mu(n-k)} \end{pmatrix}.$$

Then

$$\mathrm{sgn}(\gamma^{\nu\mu}) = \mathrm{sgn}(\gamma)\mathrm{sgn}(\nu)\mathrm{sgn}(\mu). \tag{3.42}$$

PROOF. Note that the second line of $\gamma^{\nu\mu}$ can be converted to the second line of γ by first transposition sorting ν to transform $(y_{\nu(1)}, \ldots, y_{\nu(k)})$ to $(y_1, \ldots, y_k)$ and then transposition sorting $(y'_{\mu(1)}, \ldots, y'_{\mu(n-k)})$ to $(y_1', \ldots, y'_{n-k})$. □

LEMMA 3.43 (**SIGN OF γ_y**). *We use the terminology of Lemma 3.41. Let Y be ordered as integers $y_1, \ldots, y_k$ and Y' ordered $y'_1, \ldots, y'_{n-k}$. Let γ_y be the permutation*

$$\gamma_y = \begin{pmatrix} 1 & \ldots & k & k+1 & \ldots & n \\ y_1 & \ldots & y_k & y'_1 & \ldots & y'_{n-k} \end{pmatrix}.$$

Then $\operatorname{sgn}(\gamma_y) = (-1)^{\sum_{i=1}^{k}(y_i - i)} = (-1)^{\sum_{i=1}^{k} y_i}(-1)^{k(k+1)/2}$.

PROOF. Induct on $k \geq 0$. The case $k = 1$ is instructive:

$$\gamma_y = \begin{pmatrix} 1 & 2 & \ldots & n \\ y_1 & y'_1 & \ldots & y'_{n-1} \end{pmatrix}.$$

Since $y'_1 < \cdots < y'_{n-1}$, we must do exactly $y_1 - 1$ transpositions of y_1 with the y'_i to get to the sequence $1, 2, \ldots, n$. Thus, for $k = 1$, $\operatorname{sgn}(\gamma_y) = (-1)^{y_1 - 1}$ which proves the lemma for $k = 1$. Next, assume the case $k - 1$ and consider

$$\gamma_y = \begin{pmatrix} 1 & \ldots & k & k+1 & \ldots & n \\ y_1 & \ldots & y_k & y'_1 & \ldots & y'_{n-k} \end{pmatrix}.$$

First, sort y_k with $y'_1 < \cdots < y'_{n-k}$. This can be done in $y_k - 1 - (k-1) = y_k - k$ transpositions since $y_1 < \cdots < y_{k-1} < y_k$ and thus y_k doesn't have to be transposed with the $k - 1$ numbers $y_1, \ldots, y_{k-1}$. Thus, we have $\operatorname{sgn}(\gamma_y) = (-1)^{(y_k - k)}\operatorname{sgn}(\gamma'_y)$ where, using terminology analogous to that for γ_y,

$$\gamma'_y = \begin{pmatrix} 1 & \ldots & k-1 & k & \ldots & n \\ y_1 & \ldots & y_{k-1} & y'_1 & \ldots & y'_{n-(k-1)} \end{pmatrix}.$$

Applying the induction hypothesis (case $k-1$) to γ'_y gives $\operatorname{sgn}(\gamma_y) = (-1)^{\sum_{i=1}^{k}(y_i - i)}$ which was to be shown. The term $(-1)^{k(k+1)/2}$ in the statement of the lemma comes from writing $\sum_{i=1}^{k}(y_i - i) = \sum_{i=1}^{k} y_i - \sum_{i=1}^{k} i$ and using the fact that $\sum_{i=1}^{k} i = k(k+1)/2$. □

REMARK 3.44 (**EXAMPLE OF COMPUTING** $\operatorname{sgn}(\gamma_y)$). As an example of Lemma 3.43, take $k = 3$ and $n = 6$. Take $(y_1, y_2, y_3) = (3, 4, 6)$ and $(y'_1, y'_2, y'_3) = (1, 2, 5)$. Thus,

$$\gamma_y = \begin{pmatrix} 1 & 2 & 3 & 4 & 5 & 6 \\ 3 & 4 & 6 & 1 & 2 & 5 \end{pmatrix}.$$

From Lemma 3.43, $\operatorname{sgn}(\gamma_y) = (-1)^{\sum_{i=1}^{k}(y_i - i)} = (-1)^{(3-1)+(4-2)+(6-3)} = -1$. In cycle form, $\gamma_y = (1, 3, 6, 5, 2, 4)$.

The next lemma computes $\operatorname{sgn}(\gamma)$ and $\operatorname{sgn}(\gamma^{\nu\mu})$. We use the terminology of Lemmas 3.41 and 3.43.

Lemma 3.45 (**Signs of γ and $\gamma^{\nu\mu}$**). *Let γ and $\gamma^{\nu\mu}$ be as in 3.41. Let X be ordered as integers $x_1, \ldots, x_k$ and X' be ordered $x'_1, \ldots, x'_{n-k}$. Similarly, let Y be ordered $y_1, \ldots, y_k$ and Y' be ordered $y'_1, \ldots, y'_{n-k}$. For simplicity, let $\sum_{i=1}^{k} x_i \equiv \sum X$ and $\sum_{i=1}^{k} y_i = \sum Y$. Then,*

$$\operatorname{sgn}(\gamma) = (-1)^{\sum X}(-1)^{\sum Y} \quad \text{and} \quad \operatorname{sgn}(\gamma^{\nu\mu}) = (-1)^{\sum X}(-1)^{\sum Y}\operatorname{sgn}(\nu)\operatorname{sgn}(\mu).$$

Proof. Recall

$$\gamma = \begin{pmatrix} x_1 & \ldots & x_k & x'_1 & \ldots & x'_{n-k} \\ y_1 & \ldots & y_k & y'_1 & \ldots & y'_{n-k} \end{pmatrix}.$$

Let

$$\gamma_y = \begin{pmatrix} 1 & \ldots & k & k+1 & \ldots & n \\ y_1 & \ldots & y_k & y'_1 & \ldots & y'_{n-k} \end{pmatrix} \tag{3.46}$$

and let

$$\gamma_x = \begin{pmatrix} 1 & \ldots & k & k+1 & \ldots & n \\ x_1 & \ldots & x_k & x'_1 & \ldots & x'_{n-k} \end{pmatrix}. \tag{3.47}$$

Note that $\gamma\,\gamma_x = \gamma_y$. Applying Lemma 3.43 to γ_x and γ_y we get

$$\operatorname{sgn}(\gamma)\,(-1)^{\sum_{i=1}^{k} x_i}(-1)^{k(k+1)/2} = (-1)^{\sum_{i=1}^{k} y_i}(-1)^{k(k+1)/2}$$

and thus $\operatorname{sgn}(\gamma) = (-1)^{\sum_{i=1}^{k} x_i}(-1)^{\sum_{i=1}^{k} y_i} = (-1)^{\sum X}(-1)^{\sum Y}$. From 3.42 we obtain

$$\operatorname{sgn}(\gamma^{\nu\mu}) = (-1)^{\sum X}(-1)^{\sum Y}\operatorname{sgn}(\nu)\operatorname{sgn}(\mu). \tag{3.48}$$

□

Theorem 3.49 (**Determinants of general direct sums**). *Let A be an $n \times n$ matrix and let $X, Y \in \mathbb{P}_k(n)$ be fixed subsets of $\{1, \ldots, n\}$ of size k. Let $\sum X = \sum_{x\in X} x$ and let X' be the complement of X in $\{1, \ldots, n\}$ (similarly for Y, Y'). If $A = B \oplus_X^Y C$ then $B = A[X \mid Y]$ and $C = A(X \mid Y)$ and*

$$\det(A) = (-1)^{\sum X}(-1)^{\sum Y} \det A[X \mid Y] \det A(X \mid Y). \tag{3.50}$$

Proof. For any $n \times n$ matrix A, we show that the *restricted determinant sum*

$$\Delta(X, Y, A) = \sum_{\sigma\in S_X^Y} \operatorname{sgn}(\sigma) \prod_{i=1}^{n} A(i, \sigma(i)) \tag{3.51}$$

(sum over S_X^Y only) satisfies

$$\Delta(X, Y, A) = (-1)^{\sum X}(-1)^{\sum Y} \det(A[X \mid Y]) \det(A(X \mid Y)). \tag{3.52}$$

If $A = B \oplus_X^Y C$ then the restricted determinant sum gives $\det(A)$ and thus proves the result. Use the characterization of S_X^Y given in Lemma 3.39:

$$S_X^Y = \{\gamma^{\nu\mu} \mid \nu \in \mathrm{PER}(k),\ \mu \in \mathrm{PER}(n-k)\}$$

where

$$(3.53) \qquad \gamma^{\nu\mu} = \begin{pmatrix} x_1 & \ldots & x_k & x'_1 & \ldots & x'_{n-k} \\ y_{\nu(1)} & \ldots & y_{\nu(k)} & y'_{\mu(1)} & \ldots & y'_{\mu(n-k)} \end{pmatrix}.$$

The restricted determinant sum (3.51) becomes

$$(3.54) \qquad \Delta(X, Y, A) = \sum_{\nu} \sum_{\mu} \operatorname{sgn}(\gamma^{\nu\mu}) \prod_{i=1}^{k} A(x_i, y_{\nu(i)}) \prod_{i=1}^{n-k} A(x'_i, y'_{\mu(i)})$$

where $\nu \in \mathrm{PER}(k)$ and $\mu \in \mathrm{PER}(n-k)$. From 3.45 we obtain

$$(3.55) \qquad \operatorname{sgn}(\gamma^{\nu\mu}) = (-1)^{\sum X}(-1)^{\sum Y}\operatorname{sgn}(\nu)\operatorname{sgn}(\mu).$$

Thus, 3.54 becomes $\Delta(X, Y, A) =$

$$\sum_{\nu}\sum_{\mu}(-1)^{\sum X}(-1)^{\sum Y}\operatorname{sgn}(\nu)\operatorname{sgn}(\mu)\prod_{i=1}^{k} A(x_i, y_{\nu(i)})\prod_{i=1}^{n-k} A(x'_i, y'_{\mu(i)}) =$$

$$(-1)^{\sum X}(-1)^{\sum Y}\left(\sum_{\nu}\operatorname{sgn}(\nu)\prod_{i=1}^{k} A(x_i, y_{\nu(i)})\right)\left(\sum_{\mu}\operatorname{sgn}(\mu)\prod_{i=1}^{n-k} A(x'_i, y'_{\mu(i)})\right) =$$

$$(-1)^{\sum X}(-1)^{\sum Y}\det(A[X \mid Y])\det(A(X \mid Y)).$$

If $A = B \oplus_X^Y C$ then $\Delta(X, Y, A) = \det(A)$ which completes the proof. □

Corollary 3.56 (**Restricted determinant sums**). *We use the terminology of Theorem 3.49. Let A be an $n \times n$ matrix. Let*

$$\Delta(X, Y, A) = \sum_{\sigma \in S_X^Y} \operatorname{sgn}(\sigma) \prod_{i=1}^{n} A(i, \sigma(i))$$

denote the restricted determinant sum (3.51). Then

$$\Delta(X, Y, A) = (-1)^{\sum X}(-1)^{\sum Y}\det(A[X \mid Y])\det(A(X \mid Y)).$$

Proof. The proof is developed in the process of proving Theorem 3.49. □

Laplace expansion theorem

We derive the general Laplace expansion theorem. Our proof is valid for matrices with entries in a commutative ring (e.g., the Euclidean domains, $\mathbb{K}$, 1.30) and is based on Corollary 3.56. We use Definitions 1.33, 1.50 and 2.52. We also use the notation discussed in Remark 1.32.

DEFINITION 3.57 (**LAPLACE PARTITION AND CANONICAL SDR**). Let S = PER(n) be the permutatons of $\underline{n}$, and let $X, Y \in \mathbb{P}_k(n)$ be subsets of $\underline{n}$ of size k. Let

$$S_X^Y = \{\sigma \mid \sigma \in S \text{ and } \sigma(X) = Y\}. \tag{3.58}$$

For a fixed X, the collection of sets

$$\mathbb{L}^X = \{S_X^Y \mid Y \in \mathbb{P}_k(n)\} \tag{3.59}$$

is the *Laplace partition* of S = PER(n) associated with X. Let X' and Y' denote the complements of X and Y in $\underline{n}$. For $\sigma \in S_X^Y$, let σ_X and $\sigma_{X'}$ be the restrictions (1.43) of σ to X and X'. For X fixed, the set

$$D^X = \{\gamma \mid \gamma_X \in \mathrm{SNC}(X, Y),\ \gamma_{X'} \in \mathrm{SNC}(X', Y'),\ Y \in \mathbb{P}_k(n)\} \tag{3.60}$$

is the *canonical system of distinct representatives (SDR)* for the Laplace partition of S associated with X.

Note that γ_X and $\gamma_{X'}$ in 3.60 are unique since $|X| = |Y|$ and $|X'| = |Y'|$. The number of blocks in the partition $\mathbb{L}^X$ is $|\mathbb{L}^X| = \binom{n}{k}$.

THEOREM 3.61 (**LAPLACE EXPANSION THEOREM**). *Let A be an $n \times n$ matrix and let $X \in \mathbb{P}_k(n)$ be a fixed subset of $\{1, \ldots, n\}$ of size k. Let $\sum X = \sum_{x \in X} x$. Then the fixed-rows form of the Laplace expansion is*

$$\det(A) = (-1)^{\sum X} \sum_{Y \in \mathbb{P}_k(n)} (-1)^{\sum Y} \det(A[X \mid Y]) \det(A(X \mid Y)) \tag{3.62}$$

and the fixed-columns form is

$$\det(A) = (-1)^{\sum X} \sum_{Y \in \mathbb{P}_k(n)} (-1)^{\sum Y} \det(A[Y \mid X]) \det(A(Y \mid X)). \tag{3.63}$$

PROOF. By definition,

$$\det(A) = \sum_{\sigma \in \mathrm{PER}(n)} \mathrm{sgn}(\sigma) \prod_{i=1}^{n} A(i, \sigma(i)). \tag{3.64}$$

Let $\mathbb{L}^X = \{S_X^Y \mid Y \in \mathbb{P}_k(n)\}$ be the Laplace partition of PER(n) (3.57). Then

$$\det(A) = \sum_{Y \in \mathbb{P}_k(n)} \sum_{\sigma \in S_X^Y} \mathrm{sgn}(\sigma) \prod_{i=1}^{n} A(i, \sigma(i)). \tag{3.65}$$

By definition 3.51, the inner sum of 3.65 is the restricted determinant sum $\Delta(X, Y, A)$. Thus,

$$\det(A) = \sum_{Y \in \mathbb{P}_k(n)} \Delta(X, Y, A). \tag{3.66}$$

By Corollary 3.56

$$\Delta(X, Y, A) = (-1)^{\sum X}(-1)^{\sum Y} \det(A[X \mid Y]) \det(A(X \mid Y)).$$

The column form 3.63 follows by replacing A by its transpose A^T (3.6). This completes the proof. □

Let X' be the complement of X in $\{1, \ldots, n\}$. Using the notation (2.55) $A[X' \mid Y'] = A(X \mid Y)$ we can write the fixed-row Laplace expansion (3.61) as

$$(3.67) \qquad \det(A) = (-1)^{\sum X} \sum_{Y \in \mathbb{P}_k(n)} (-1)^{\sum Y} \det(A[X \mid Y]) \det(A[X \mid Y'])$$

and the fixed-column form as

$$(3.68) \qquad \det(A) = (-1)^{\sum X} \sum_{Y \in \mathbb{P}_k(n)} (-1)^{\sum Y} \det(A[Y \mid X]) \det(A[Y' \mid X']).$$

The set of subsets of $\underline{n}$ of size k, $\mathbb{P}_k(n)$, corresponds bijectively to the set, $\mathrm{SNC}(k, n)$, of strictly increasing functions from $\underline{k}$ to $\underline{n}$. The natural bijection is for $Y \in \mathbb{P}_k(n)$, $Y = \{y_1, \ldots, y_k\}$ with $y_1 < \cdots < y_k$, to correspond to $\begin{pmatrix} 1 & \ldots & k \\ y_1 & \ldots & y_k \end{pmatrix}$. Thus, we can rewrite the Laplace expansion theorem in terms of functions. We use notation like that of 2.55. Let $g \in \mathrm{SNC}(k, n)$ be fixed, let $A = (a_{ij})$ be an $n \times n$ matrix and let $\sum g = \sum_{i=1}^{k} g(i)$. Then the fixed row Laplace expansion expressed in terms of functions is

$$(3.69) \qquad \det(A) = (-1)^{\sum g} \sum_{f \in \mathrm{SNC}(k,n)} (-1)^{\sum f} \det(A[g \mid f]) \det(A(g \mid f))$$

and the fixed-column Laplace expansion expressed in terms of functions is

$$(3.70) \qquad \det(A) = (-1)^{\sum g} \sum_{f \in \mathrm{SNC}(k,n)} (-1)^{\sum f} \det(A[f \mid g] \det(A(f \mid g)).$$

REMARK 3.71 (**EXAMPLE OF LAPLACE EXPANSION**). Let A be a 4×4 matrix of integers ($A \in \mathbf{M}_{4,4}(\mathbb{Z})$)

$$A = \begin{pmatrix} 1 & 2 & 3 & 4 \\ 2 & 3 & 4 & 1 \\ 3 & 4 & 1 & 2 \\ 4 & 1 & 2 & 3 \end{pmatrix}.$$

Use the fixed-row Laplace expansion (3.69) with the fixed function $g = (1, 3)$ (in one line notation). Take the variable functions, f, lexicographically in one-line notation: $(1, 2), (1, 3), (1, 4), (2, 3), (2, 4), (3, 4)$. Then $\det(A) \equiv |A| =$

$$-\begin{vmatrix} 1 & 2 \\ 3 & 4 \end{vmatrix}\begin{vmatrix} 4 & 1 \\ 2 & 3 \end{vmatrix} + \begin{vmatrix} 1 & 3 \\ 3 & 1 \end{vmatrix}\begin{vmatrix} 3 & 1 \\ 1 & 3 \end{vmatrix} - \begin{vmatrix} 1 & 4 \\ 3 & 2 \end{vmatrix}\begin{vmatrix} 3 & 4 \\ 1 & 2 \end{vmatrix}$$

$$-\begin{vmatrix} 2 & 3 \\ 4 & 1 \end{vmatrix}\begin{vmatrix} 2 & 1 \\ 4 & 3 \end{vmatrix} + \begin{vmatrix} 2 & 4 \\ 4 & 2 \end{vmatrix}\begin{vmatrix} 2 & 4 \\ 4 & 2 \end{vmatrix} - \begin{vmatrix} 3 & 4 \\ 1 & 2 \end{vmatrix}\begin{vmatrix} 2 & 3 \\ 4 & 1 \end{vmatrix} = 160.$$

The next corollary is the version of the Laplace expansion theorem that is most often stated and proved in elementary courses:

Corollary 3.72 (**Simple Laplace expansion**). *Let A be an $n \times n$ matrix, $n > 1$, and let $1 \leq i \leq n$. Then*

$$\det(A) = (-1)^i \sum_{j=1}^{n} (-1)^j a_{ij} \det(A(i \mid j)) \tag{3.73}$$

$$\det(A) = (-1)^i \sum_{j=1}^{n} (-1)^j a_{ji} \det(A(j \mid i)). \tag{3.74}$$

Proof. We use 3.67:

$$\det(A) = (-1)^{\sum X} \sum_{Y \in \mathbb{P}_k(n)} (-1)^{\sum Y} \det(A[X \mid Y]) \det(A(X \mid Y)).$$

In this case,

$$\sum X = i, \ \sum Y = j, \ A[X \mid Y] = A[i \mid j] = a_{ij}, \ \text{and} \ \det(A[X \mid Y]) = a_{ij}.$$

Substituting these values and noting that $A(X \mid Y) = A(i \mid j)$ proves 3.73. Identity 3.74 follows by using 3.68 instead of 3.67. □

Note that for $A = (a_{ij})$ then $A(i, j)$ or $A[i \mid j]$ can be used in place of a_{ij} in 3.73. Recall the "delta" notation: δ(**statement**) = 0 if **statement** is false, 1 if **statement** is true.

Corollary 3.75 (**Simple Laplace extended**). *Let A be an $n \times n$ matrix, $n > 1$, and let $1 \leq i \leq n$. Then*

$$\delta(k = i) \det(A) = (-1)^k \sum_{j=1}^{n} (-1)^j a_{ij} \det(A(k \mid j)). \tag{3.76}$$

or, alternatively,

$$AB_A(i, k) = \sum_{j=1}^{n} A(i, j) B_A(j, k) = \delta(i = k) \det(A) \tag{3.77}$$

where $B_A = (b_{jk})$ and $b_{jk} = (-1)^{(k+j)} \det(A(k \mid j))$.

Proof. The case where $k = i$ becomes

$$\det(A) = (-1)^i \sum_{j=1}^{n} (-1)^j a_{ij} \det(A(i \mid j)) \tag{3.78}$$

which is a simple Laplace expansion (3.73).

Consider $k \neq i$. Take the matrix A with rows

$$A = (A_{(1)}, \ldots, A_{(k-1)}, A_{(k)}, A_{(k+1)}, \ldots, A_{(n)})$$

and replace row k, $A_{(k)}$, with row i, $A_{(i)}$, to obtain a matrix A':

$$A' = (A_{(1)}, \ldots, A_{(k-1)}, A_{(i)}, A_{(k+1)}, \ldots, A_{(n)}).$$

The matrix A' thus has two identical rows and hence $\det(A') = 0$.

Apply equation 3.73 to A' to get

$$\det(A') = (-1)^k \sum_{j=1}^{n} (-1)^j A'(k,j) \det(A'(k \mid j)). \tag{3.79}$$

By definition of A', $A'(k,j) = A(i,j) = a_{ij}$. Also by definition of A', the $(n-1) \times (n-1)$ matrix $A'(k \mid j) = A(k \mid j)$. Thus by 3.79 we have

$$0 = \det(A') = (-1)^k \sum_{j=1}^{n} (-1)^j a_{ij} \det(A(k \mid j)) \quad \text{where} \quad k \neq i. \tag{3.80}$$

Combining 3.78 and 3.80 gives 3.76.

Rewrite equation 3.76 as follows:

$$\sum_{j=1}^{n} a_{ij}[(-1)^{k+j} \det(A(k \mid j))] = \delta(i = k) \det(A). \tag{3.81}$$

If we define a matrix $B_A = (b_{jk})$ by $b_{jk} = (-1)^{(k+j)} \det(A(k \mid j))$ then we obtain

$$AB_A(i,k) = \sum_{j=1}^{n} A(i,j) B_A(j,k) = \delta(i = k) \det(A). \tag{3.82}$$

This completes the proof. □

Definition 3.83 (**Signed cofactor matrix**). Let A be an $n \times n$ matrix. For $1 \leq i,j \leq n$, define $c_{ij} = (-1)^{i+j} \det A(i \mid j)$. We call c_{ij} the *signed cofactor* of $A(i,j)$. The $n \times n$ matrix $C_A = (c_{ij})$ is the *signed cofactor matrix* of A. The transpose, $B_A = C_A^T$, of C_A is sometimes called the *adjugate* of A and written $\mathrm{adj}(A)$.

The next corollary restates 3.82. We use $\det(XY) = \det(X)\det(Y)$ (3.99).

Corollary 3.84. *Let $C_A = (c_{ij})$ be the signed cofactor matrix of A. Let $B_A = C^T$ be the transpose of C_A (i.e., $B_A = \mathrm{adj}(A)$) . Then*

$$AB_A = B_A A = (\det A) I_n \tag{3.85}$$

where I_n is the $n \times n$ identity matrix. If $\det(A)$ is a unit in $\mathbb{K}$ then A is a unit in $\mathbf{M}_{n,n}(\mathbb{K})$ (1.30) and $A^{-1} = B_A(\det(A))^{-1}$. Thus, A is unit in $\mathbf{M}_{n,n}(\mathbb{K})$ if and only if $\det(A)$ is a unit in $\mathbb{K}$.

Proof. If $\det(A) \neq 0$ then $A^{-1} = B_A(\det(A))^{-1}$ follows from 3.82. The converse follows from the fact that if $AA^{-1} = I_n$ then $\det(A)\det(A^{-1}) = 1$ so $(\det(A))^{-1} = \det(A^{-1})$. If $\det(A) \neq 0$ then $AB_A = B_A A$ follows from the fact that $A^{-1}A = AA^{-1}$. The statement $AB_A = (\det A) I_n$ is exactly the same as equation 3.82 and does not require A to be nonsingular. In general, commutivity, $AB_A = B_AA$, follows from 3.75 by replacing A by A^T in equation 3.82:

$$\delta(k = i)\det(A^T) = \sum_{j=1}^{n}(-1)^{k+j}a^T_{ij}\det(A^T(k \mid j)). \tag{3.86}$$

$\det(A^T(k \mid j)) = \det((A(j \mid k))^T) = \det(A(j \mid k))$ (3.7) and

$$(-1)^{k+j}\det(A(j \mid k)) = B_A(k, j)$$

thus using 3.86

$$\delta(k = i)\det(A) = \sum_{j=1}^{n} B_A(k, j)a_{ji} = B_AA. \tag{3.87}$$

Thus, $AB_A = B_AA$ in all cases.

□

Remark 3.88 (**Example of signed cofactor and adjugate matrices**). Let A be a 4×4 matrix of integers ($A \in M_{4,4}(\mathbb{Z})$) with $A(i, j) = \delta(i \leq j)$. The matrices A, C_A, and $B_A = \mathrm{adj}(A)$ of Corollary 3.84 are as follows:

$$A = \begin{pmatrix} 1 & 1 & 1 & 1 \\ 0 & 1 & 1 & 1 \\ 0 & 0 & 1 & 1 \\ 0 & 0 & 0 & 1 \end{pmatrix} \quad C_A = \begin{pmatrix} +1 & 0 & 0 & 0 \\ -1 & +1 & 0 & 0 \\ 0 & -1 & +1 & 0 \\ 0 & 0 & -1 & +1 \end{pmatrix} \quad B_A = \begin{pmatrix} +1 & -1 & 0 & 0 \\ 0 & +1 & -1 & 0 \\ 0 & 0 & +1 & -1 \\ 0 & 0 & 0 & +1 \end{pmatrix}.$$

If A is an $n \times n$ matrix over a field $\mathbb{F}$ and A^{-1} is the inverse of A, then $\det(AA^{-1}) = \det(A)\det(A^{-1}) = \det(I_n) = 1$. If X is $n \times 1$, we write $X(t) = X(t, 1)$, $t = 1, \ldots, n$. Putting together Definition 3.83 and Corollary 3.84 we get the following corollary.

Corollary 3.89 (**Cramer's rule**). *Let $AX = Y$ where A is $n \times n$, X is $n \times 1$ and Y is $n \times 1$ (entries in a field $\mathbb{F}$). Designate $A = (A^{(1)}, \ldots, A^{(i)}, \ldots, A^{(n)})$ as a sequence of columns. Define $\hat{A} = (A^{(1)}, \ldots, A^{(i-1)}, Y, A^{(i+1)}, \ldots, A^{(n)})$ to be the matrix A with column $A^{(i)}$ replaced by Y. Then*

$$X(i) = \det(\hat{A})/\det(A).$$

Proof. $AX = Y$ implies $X = A^{-1}Y$ and $X(i) = \sum_{j=1}^{n} A^{-1}(i, j)Y(j)$. From Definition 3.83 and Corollary 3.84 we get

$$A^{-1}(i, j) = \left[(-1)^{i+j}\frac{\det(A(j \mid i))}{\det(A)}\right]. \tag{3.90}$$

Thus

$$X(i) = \sum_{j=1}^{n} A^{-1}(i, j)Y(j) = \sum_{j=1}^{n}\left[(-1)^{i+j}\frac{\det(A(j \mid i))}{\det(A)}\right]Y(j) \tag{3.91}$$

and

$$X(i) = \frac{1}{\det(A)} \sum_{j=1}^{n} (-1)^{i+j} \det(A(j \mid i)) Y(j). \tag{3.92}$$

Observe that the matrix $A(j \mid i) = \hat{A}(j \mid i)$ since only columns i differ between A and $\hat{A}$. Note also that $Y(j) = \hat{A}(j, i)$. Thus, the sum of 3.92 becomes

$$\sum_{j=1}^{n} (-1)^{i+j} \det(A(j \mid i)) Y(j) = (-1)^{i} \sum_{j=1}^{n} (-1)^{j} \hat{A}(j, i) \det(\hat{A}(j \mid i)) = \det(\hat{A})$$

using 3.72 (column form, 3.74). This completes the proof. □

Remark 3.93 (**Example of Cramer's rule**). The two equations

$$\begin{array}{ccccc} x_1 & + & x_2 & = & 3 \\ x_1 & - & x_2 & = & 1 \end{array}$$

can be expressed by the equation $AX = Y$ where

$$A = \begin{pmatrix} 1 & 1 \\ 1 & -1 \end{pmatrix} \quad X = \begin{pmatrix} x_1 \\ x_2 \end{pmatrix} \quad Y = \begin{pmatrix} 3 \\ 1 \end{pmatrix}.$$

Applying 3.89 twice, to $X(1) = x_1$ and $X(2) = x_2$, and noting that $\det(A) = -2$ gives

$$x_1 = \frac{\det\begin{pmatrix} 3 & 1 \\ 1 & -1 \end{pmatrix}}{-2} = 2 \quad \text{and} \quad x_2 = \frac{\det\begin{pmatrix} 1 & 3 \\ 1 & 1 \end{pmatrix}}{-2} = 1.$$

Cauchy-Binet theorem

We need some notational conventions for describing product - sum interchanges. Consider the following example:

$$(x_{11} + x_{12})(x_{21} + x_{22}) = x_{11}x_{21} + x_{11}x_{22} + x_{12}x_{21} + x_{12}x_{22}.$$

Look at the second integers in each pair of subscripts:

$$x_{1\underline{1}}x_{2\underline{1}} + x_{1\underline{1}}x_{2\underline{2}} + x_{1\underline{2}}x_{2\underline{1}} + x_{1\underline{2}}x_{2\underline{2}}.$$

The pairs of underlined integers are, in order:

$$11, 12, 21, 22.$$

These pairs represent (in one line notation) all of the functions in $\underline{2}^{\underline{2}}$. Thus, we can write

$$\prod_{i=1}^{2} \left(\sum_{k=1}^{2} x_{ik} \right) = \sum_{f \in \underline{2}^{\underline{2}}} \prod_{i=1}^{2} x_{i f(i)}. \tag{3.94}$$

The general form of this identity is

$$\prod_{i=1}^{n}\left(\sum_{k=1}^{p} x_{ik}\right) = \sum_{f\in \underline{p}^{\underline{n}}} \prod_{i=1}^{n} x_{if(i)}. \tag{3.95}$$

This product-sum-interchange identity 3.95 is important to what follows.

We now prove the Cauchy-Binet theorem using 3.19, 2.52, 3.95 and 1.51. The proof is valid for matrices with entries in a commutative ring (e.g., a Euclidean domain).

THEOREM 3.96 (**Cauchy-Binet**). *Let A be an $n \times p$ and B a $p \times n$ matrix. Then*

$$\det(AB) = \sum_{f\in \mathrm{SNC}(n,p)} \det(A^f)\det(B_f)$$

where $\mathrm{SNC}(n,p)$ *denotes the strictly increasing functions from* $\underline{n}$ *to* $\underline{p}$ *(see 1.50);* A^f *denotes the submatrix of* A *with columns selected by* f*; and* B_f *denotes the submatrix of* B *with rows selected by* f *(see 2.52).*

PROOF.

$$\det(AB) = \sum_{\gamma\in \mathrm{PER}(n)} \mathrm{sgn}(\gamma) \prod_{i=1}^{n} (AB)(i,\gamma(i)) \quad \textbf{(definition } \det\textbf{)}$$

$$\det(AB) = \sum_{\gamma\in \mathrm{PER}(n)} \mathrm{sgn}(\gamma) \prod_{i=1}^{n}\sum_{k=1}^{p} A(i,k)B(k,\gamma(i)) \quad \textbf{(definition } AB\textbf{)}$$

$$\det(AB) = \sum_{\gamma\in \mathrm{PER}(n)} \mathrm{sgn}(\gamma) \sum_{h\in \underline{p}^{\underline{n}}} \prod_{i=1}^{n} A(i,h(i))B(h(i),\gamma(i)) \quad \textbf{(identity } 3.95\textbf{)}$$

$$\det(AB) = \sum_{h\in \underline{p}^{\underline{n}}} \prod_{i=1}^{n} A(i,h(i)) \sum_{\gamma\in \mathrm{PER}(n)} \mathrm{sgn}(\gamma) \prod_{i=1}^{n} B(h(i),\gamma(i)) \quad \textbf{(algebra rules)}$$

$$\det(AB) = \sum_{h\in \underline{p}^{\underline{n}}} \prod_{i=1}^{n} A(i,h(i))\det(B_h) \quad \textbf{(definition } B_h \ 2.52\textbf{)} \tag{3.97}$$

Now observe that $\det(B_h) = 0$ if $h \notin \mathrm{INJ}(n,p)$ by 3.20. Thus,

$$\det(AB) = \sum_{h\in \mathrm{INJ}(n,p)} \prod_{i=1}^{n} A(i,h(i))\det(B_h) \quad \textbf{(determinant property } 3.17\textbf{)}.$$

Thus,

$$\det(AB) = \sum_{f\in \mathrm{SNC}(n,p)} \sum_{\gamma\in \mathrm{PER}(n)} \left(\prod_{i=1}^{n} A(i,f\gamma(i))\right)\det(B_{f\gamma}) \quad \textbf{(set identity } 1.51\textbf{)}$$

Note that $A(i, f\gamma(i)) = A^f(i, \gamma(i))$. The matrix $B_{f\gamma}$ of the previous equation is the same as $(B_f)_\gamma$. Identity 3.19 implies that $\det((B_f)_\gamma) = \operatorname{sgn}(\gamma)\det(B_f)$. Thus we obtain

$$\det(AB) = \sum_{f\in\mathrm{SNC}(n,p)} \left(\sum_{\gamma\in\mathrm{PER}(n)} \operatorname{sgn}(\gamma) \prod_{i=1}^{n} A^f(i,\gamma(i)) \right) \det(B_f). \quad \textbf{(algebra rules)}$$

Finally, applying the definition of the determinant to A^f we get

$$\det(AB) = \sum_{f\in\mathrm{SNC}(n,p)} \det(A^f)\det(B_f).$$

□

Remark 3.98 (**Zero determinant of product**). Equation 3.97 of the preceding theorem is a sum of the form

$$\det(AB) = \sum_{h\in \underline{p}^{\underline{n}}} C(h)\det(B_h)$$

where $C(h)$ depends on A and h. If $n > p$ then every $h \in \underline{p}^{\underline{n}}$ has $h(s) = h(t)$ for some pair of values $s < t$ (i.e., the set $\mathrm{INJ}(n,p)$ is empty). For every such h, $\det(B_h) = 0$ and hence $\det(AB) = 0$ if $n > p$.

Corollary 3.99 (**Determinant of product**). *If A and B are $n \times n$ matrices, then* $\det(AB) = \det(A)\det(B)$.

Proof. Apply the Cauchy-Binet theorem (3.96) with $p = n$. In that case, $\det(AB) = \sum_{f\in\mathrm{SNC}(n,n)} \det(A^f)\det(B_f)$. $\mathrm{SNC}(n,n)$ has only one element, the identity function $f(i) = i$ for all $i \in \underline{n}$. Thus, $A^f = A$ and $B_f = B$. □

Corollary 3.100 is a useful restatement of Theorem 3.96.

Corollary 3.100 (**General Cauchy-Binet**). *Let A be an $a \times p$ matrix and B a $p \times b$ matrix. Let $g \in \underline{a}^{\underline{n}}$ and $h \in \underline{b}^{\underline{n}}$. Let $C = AB$. Then*

$$\det(C[g \mid h]) = \sum_{f\in\mathrm{SNC}(n,p)} \det(A[g \mid f])\det(B[f \mid h]) \tag{3.101}$$

Proof. We apply Theorem 3.96 to A_g $(n \times p)$ and B^h $(p \times n)$:

$$\det(A_g B^h) = \sum_{f\in\mathrm{SNC}(n,p)} \det((A_g)^f)\det((B^h)_f).$$

This becomes

$$\det((AB)[g \mid h]) = \sum_{f\in\mathrm{SNC}(n,p)} \det(A[g \mid f])\det(B[f \mid h]).$$

Substituting $C = AB$ gives 3.101.

□

Remark 3.102 (**General Cauchy-Binet**). If $n > p$ then $\mathrm{SNC}(n,p) = \emptyset$ (empty set). Thus, $n \le p$ is the more interesting case of 3.101. Likewise, if $n > a$ or $n > b$ then the right hand side of 3.101 is zero. Thus, we are most interested in the case $n \le \min(a,b,p)$. Even if $n \le \min(a,b,p)$ then we still need g and h to be injections to make the corresponding determinants nonzero. These observations lead to the following version:

Corollary 3.103 (**Extended Cauchy-Binet**). *Let A be an $a \times p$ matrix and B a $p \times b$ matrix. Let $g \in \underline{a}^{\underline{n}}$ and $h \in \underline{b}^{\underline{n}}$. Assume $n \le \min(a,b,p)$ and g and h are injective (1.50). Let $C = AB$. Then*

$$\det(C[g \mid h]) = \sum_{f \in \mathrm{SNC}(n,p)} \det(A[g \mid f]) \det(B[f \mid h]) \tag{3.104}$$

$$\det(C_g^h) = \sum_{f \in \mathrm{SNC}(n,p)} \det(A_g^f) \det(B_f^h) \quad \left(C_g^h = (AB)_g^h = A_g B^h\right). \tag{3.105}$$

Proof. These statements are a special case of 3.100.

□

Remark 3.106 (**g and h strictly increasing**). In Corollary 3.103 the additional assumption that g and h are strictly increasing is often made: $g \in \mathrm{SNC}(n,a)$ and $h \in \mathrm{SNC}(n,b)$ (1.50). This assumption implies the standard "set" version of the Cauchy-Binet theorem. We identify a function, $g \in \mathrm{SNC}(n,a)$, with the set $G = \mathrm{image(g)}$.

Corollary 3.107 (**Cauchy-Binet set version**). *Let A be an $a \times p$ matrix and B a $p \times b$ matrix. Let $C = AB$. Assume $n \le \min(a,b,p)$. Let $G \in \mathbb{P}_n(\underline{a})$ and $H \in \mathbb{P}_n(\underline{b})$ be subsets of size n. Then*

$$\det(C[G \mid H]) = \sum_{F \in \mathbb{P}_n(\underline{p})} \det(A[G \mid F]) \det(B[F \mid H]). \tag{3.108}$$

Alternatively, let $g \in \mathrm{SNC}(n,a)$ and $h \in \mathrm{SNC}(n,b)$ be defined by $\mathrm{image}(g) = G$ *and* $\mathrm{image}(h) = H$. *Then*

$$\det(A_g B^h) = \sum_{f \in \mathrm{SNC}(n,p)} \det(A_g^f) \det(B_f^h) \tag{3.109}$$

where A_g is $n \times p$, B^h is $p \times n$ while A_g^f and B_f^h are both $n \times n$.

Proof. This statement is a special case of 3.103. We use 2.60. We also use equation 2.61 to select rows and columns of matrices:

$$C[g \mid h] \equiv C_g^h \equiv (AB)_g^h = A_g B^h. \tag{3.110}$$

□

REMARK 3.111 (**DISCUSSION OF THEOREM** 3.96). Take A to be a 2×6 matrix (i.e. $n = 2$ and $p = 6$), and B to be a 6×2 matrix as follows:

$$(3.112)\qquad A = \begin{array}{c} \\ 1 \\ 2 \end{array}\begin{array}{c} \begin{array}{cccccc} 1 & 2 & 3 & 4 & 5 & 6 \end{array} \\ \begin{bmatrix} 2 & 5 & 4 & -2 & 2 & 1 \\ 2 & 6 & 5 & 0 & -1 & 0 \end{bmatrix} \end{array} \qquad B = \begin{array}{c} \\ 1 \\ 2 \\ 3 \\ 4 \\ 5 \\ 6 \end{array}\begin{array}{c} \begin{array}{cc} 1 & 2 \end{array} \\ \begin{bmatrix} 2 & 5 \\ 1 & 1 \\ 2 & 3 \\ 2 & 1 \\ 3 & 1 \\ 2 & 2 \end{bmatrix} \end{array}.$$

Note that $C = AB$ is a 2×2 matrix and thus $\det(C)$ is defined. However, $\det(A)$ and $\det(B)$ are not defined. We have (by 3.96)

$$(3.113)\qquad \det(AB) = \sum_{f \in \mathrm{SNC}(2,6)} \det(A^f)\det(B_f)$$

where SNC(2, 6) denotes the strictly increasing functions from $\underline{2}$ to $\underline{6}$ (see 1.50); A^f denotes the submatrix of A with columns selected by f, and B_f denotes the submatrix of B with rows selected by f. For example, take $f = (1, 3)$ in one-line notation. Then

$$A^f = A^{(1,3)} = \begin{bmatrix} 2 & 4 \\ 2 & 5 \end{bmatrix} \quad \text{and;} \quad B_f = B_{(1,3)} = \begin{bmatrix} 2 & 5 \\ 2 & 3 \end{bmatrix}.$$

Thus, $\det(A^f)\det(B_f) = 2 \cdot (-4) = -8$ is one of 15 terms in the sum of 3.113.

REMARK 3.114 (**DISCUSSION OF** 3.107). Take A to be a 4×6 ($a \times p$) and B to be a 6×3 ($p \times b$) matrix as follows:

$$(3.115)\qquad A = \begin{array}{c} \\ 1 \\ 2 \\ 3 \\ 4 \end{array}\begin{array}{c} \begin{array}{cccccc} 1 & 2 & 3 & 4 & 5 & 6 \end{array} \\ \begin{bmatrix} 2 & 5 & 4 & -2 & 2 & 1 \\ 0 & 1 & 1 & 0 & -1 & 0 \\ 2 & 6 & 5 & 0 & -1 & 0 \\ 2 & 1 & 1 & -1 & -1 & 0 \end{bmatrix} \end{array} \qquad B = \begin{array}{c} \\ 1 \\ 2 \\ 3 \\ 4 \\ 5 \\ 6 \end{array}\begin{array}{c} \begin{array}{ccc} 1 & 2 & 3 \end{array} \\ \begin{bmatrix} 2 & 5 & 5 \\ 1 & 1 & 1 \\ 2 & 3 & 5 \\ 2 & 1 & 4 \\ 3 & 1 & 1 \\ 2 & 2 & 5 \end{bmatrix} \end{array}.$$

Note that $C = AB$ is a 4×3 matrix and thus $\det(C)$ is not defined. From 3.108

$$(3.116)\qquad \det(C[G \mid H]) = \sum_{F \in \mathbb{P}_n(\underline{p})} \det(A[G \mid F])\det(B[F \mid H]).$$

We choose $n \le \min(a, b, p) = \min(4, 3, 6) = 3$ to be $n = 2$. Choose $G \in \mathbb{P}_2(\underline{4})$ to be $G = \{1, 3\}$ and $H \in \mathbb{P}_2(\underline{3})$ to be $H = \{1, 2\}$.

$$\det(C[\{1,3\} \mid \{1,2\}]) = \sum_{F \in \mathbb{P}_2(\underline{6})} \det(A[\{1,3\} \mid F])\det(B[F \mid \{1,2\}]).$$

We can rewrite this equation using 3.110.

$$\det(A_g B^h) = \sum_{f \in \mathrm{SNC}(2,6)} \det(A_g^f) \det(B_f^h) \tag{3.117}$$

where $g = (1, 3)$ and $h = (1, 2)$ in one-line notation. Note that A_g is the matrix A and B^h is the matrix B of 3.112. From 2.61 we have $A_g B^h = (AB)_g^h = C_g^h$. Thus, the matrices of Corollary 3.107 are "containers" for many instances where Theorem 3.96 can be applied.

(3.118) Rank of a matrix

Definition 3.119 (**Rank of a matrix**). Recall the notation for sets of subsets, $\mathbb{P}_n(\underline{a})$ (1.32). Let $C \in \mathbf{M}_{a,b}(\mathbb{K})$ be an $a \times b$ matrix. The *rank* $\rho(C)$ is the size of the largest nonzero sub-determinant of C:

$$\rho(C) = \max\{n \mid n \in \mathbb{N}_0,\ G \in \mathbb{P}_n(\underline{a}),\ H \in \mathbb{P}_n(\underline{b})\ \ \det(C[G \mid H])) \neq 0\}.$$

If $C = \Theta_{a,b}$ is the zero matrix, then $\rho(C) = 0$.

Remark 3.120 (**Alternative definitions of rank**). The notion of "rank" for modules was discussed previously (1.26,1.28). Definition 3.119 defines rank for a matrix. You will recall from your linear algebra courses that the rank of a matrix $C \in \mathbf{M}_{a,b}(\mathbb{K})$, $\mathbb{K}$ a field, is the same as the dimension of the row space of C which is the same as the dimension of the column space of C. This dimension is equal to the maximum number of linearly independent rows or columns of C. The following is a technically another corollary of theorem 3.96.

Corollary 3.121 (**Rank of a product**). *Let $A \in \mathbf{M}_{a,p}(\mathbb{K})$ and $B \in \mathbf{M}_{p,b}(\mathbb{K})$. Let $C = AB$. Then the rank $\rho(C)$ satisfies*

$$\rho(C) = \rho(AB) \leq \min\{\rho(A), \rho(B)\}. \tag{3.122}$$

If B is nonsingular then $\rho(AB) = \rho(A)$ and if A is nonsingular $\rho(AB) = \rho(B)$.

Proof. Let $r = \rho(AB)$. From the definition of rank, there exists $G \in \mathbb{P}_r(\underline{a})$ and $H \in \mathbb{P}_r(\underline{b})$ such that $\det((AB)[G \mid H]) \neq 0$. From 3.108,

$$\det((AB)[G \mid H]) = \sum_{F \in \mathbb{P}_r(\underline{p})} \det(A[G \mid F]) \det(B[F \mid H]) \neq 0. \tag{3.123}$$

To be able to choose the subsets F, G, and H, we have $r \leq \min(a, b, p)$. Suppose, without loss of generality, that $\rho(A) < \rho(AB) = r$. Then we have $\det(A[G \mid F]) = 0$ for every term in the sum and hence $\det((AB)[G \mid H]) = 0$, contrary to assumption. Thus, $\rho(A) \geq \rho(AB)$. Similarly, $\rho(B) \geq \rho(AB)$. Let $C = AB$. We have shown that $\rho(C) \leq \rho(B)$ whether or not A is nonsingular. If A is nonsingular, let $B = A^{-1}C$ and apply 3.122 again to get $\rho(B) \leq \rho(C)$. Thus, $\rho(B) = \rho(C)$ if A is nonsingular. The argument to show $\rho(A) = \rho(C)$ if B is nonsingular is the same.

□

REMARK 3.124 (**FUNCTION NOTATION FOR 3.121 PROOF**). The set notation for submatrices used in the proof of 3.121 is standard in the literature. The equivalent "function notation" is, however, more expressive of what is going on. Let $A \in \mathbf{M}_{a,p}(\mathbb{K})$ and $B \in \mathbf{M}_{p,b}(\mathbb{K})$ and let $r = \rho(AB)$. Let $g \in \text{SNC}(r, a)$ and $h \in \text{SNC}(r, b)$ be strictly increasing functions such that $\det((AB)_g^h) \neq 0$. Note that the $r \times r$ matrix $(AB)_g^h = A_g B^h$ where A_g is $r \times p$ and B^h is $p \times r$. Analogous to 3.109 and 3.117 we can write

$$\det((AB)_g^h) = \det(A_g B^h) = \sum_{f \in \text{SNC}(r,p)} \det(A_g^f)\det(B_f^h) \tag{3.125}$$

and use this identity instead of 3.123 in the proof of Corollary 3.121.

Exercises: Cauchy Binet and Laplace

DEFINITION 3.126 (**GREATEST COMMON DIVISOR**). Let $S \subset \mathbb{Z}$ be a finite set of integers containing at least one nonzero integer. The set of greatest common divisors of S is $\{-d, d\}$ where d is the largest positive integer that divides all of the integers in S. We call d *the* greatest common divisor, $d = \gcd(S)$. See 1.31.

EXERCISE 3.127. Let $A \in \mathbf{M}_{a,p}(\mathbb{Z})$ and $B \in \mathbf{M}_{p,b}(\mathbb{Z})$. Let $C = AB$. Assume $n \le \min(a, b, p)$. Suppose that $\det(C[G \mid H]) \neq 0$ for some $G \in \mathbb{P}_n(\underline{a})$, $H \in \mathbb{P}_n(\underline{b})$. Let

$$C_n = \gcd(\{\det(C[G \mid H]) \mid G \in \mathbb{P}_n(\underline{a}),\ H \in \mathbb{P}_n(\underline{b})\})$$
$$\mathcal{A}_n = \gcd(\{\det(A[G \mid F]) \mid G \in \mathbb{P}_n(\underline{a}),\ F \in \mathbb{P}_n(\underline{p})\})$$
$$\mathcal{B}_n = \gcd(\{\det(B[F \mid H]) \mid F \in \mathbb{P}_n(\underline{p}),\ H \in \mathbb{P}_n(\underline{b})\})$$

where gcd denotes the greatest common divisor. Prove that $\det(A[G \mid F]) \neq 0$ for some G and F, $\det(B[F \mid H]) \neq 0$ for some F and H and $\mathcal{A}_n$ and $\mathcal{B}_n$ both divide C_n (recall 3.107).

EXERCISE 3.128. Repeat example 3.88 with

$$A = \begin{pmatrix} 1 & 1 & 1 \\ -1 & -2 & -2 \\ 1 & 2 & 1 \end{pmatrix}$$

by finding C_A, the signed cofactor matrix, and $B_A = C_A^T$, the transpose of C_A. Verify the identity of 3.84. Is this matrix, A, an invertible element (i.e., a unit 1.16) in the ring $\mathbf{M}_{3,3}(\mathbb{Z})$?

EXERCISE 3.129. Use Cramer's rule (3.89) to find X in the equation $AX = Y$ where

$$A = \begin{pmatrix} 1 & 0 & 1 \\ -1 & 1 & 0 \\ 0 & -1 & 2 \end{pmatrix} \quad X = \begin{pmatrix} x_1 \\ x_2 \\ x_3 \end{pmatrix} \quad Y = \begin{pmatrix} 1 \\ 0 \\ 1 \end{pmatrix}.$$

Lemma 4.6. *Let $A, B \in \mathbf{M}_{n,m}(\mathbb{K})$ be left-unit equivalent as in Definition 4.2. Thus, $A = QB$, $Q \in \mathbf{M}_{n,n}(\mathbb{K})$ a unit. Let $(A_{(1)}, \ldots, A_{(n)})$ and $(B_{(1)}, \ldots, B_{(n)})$ be the sequences of row vectors of A and B. Then*

$$\mathrm{Span}((A_{(1)}, \ldots, A_{(n)}) = \mathrm{Span}((B_{(1)}, \ldots, B_{(n)}).$$

Proof. Let $\sum_{i=1}^{n} a_i A_{(i)}$ be a linear combination of the rows of A. We show that there is a linear combination of the rows of B such that

$$\sum_{i=1}^{n} b_i B_{(i)} = \sum_{i=1}^{n} a_i A_{(i)}.$$

In matrix terms, $(b_1, \ldots, b_n)B = (a_1, \ldots, a_n)A$. This latter identity can be written

$$(b_1, \ldots, b_n)(QA) = (a_1, \ldots, a_n)A$$

which can be solved by taking

$$(b_1, \ldots, b_n) = (a_1, \ldots, a_n)Q^{-1}.$$

Thus,

$$\mathrm{Span}((A_{(1)}, \ldots, A_{(n)}) \subseteq \mathrm{Span}((B_{(1)}, \ldots, B_{(n)}).$$

The reverse inclusion follows from $B = Q^{-1}A$ and the same argument. □

We now define elementary row and column operations on a matrix. Let $\mathbb{K}$ be as in (1.30).

Definition 4.7 (**Elementary row and column operations**). Let $A \in \mathbf{M}_{n,m}(\mathbb{K})$ be an $n \times m$ matrix. Define three types of functions from $\mathbf{M}_{n,m}(\mathbb{K})$ to $\mathbf{M}_{n,m}(\mathbb{K})$ called *elementary row operations*:

(Type I) $\hat{R}_{[i][j]}(A)$ interchanges row $A_{(i)}$ with row $A_{(j)}$.

(Type II) $\hat{R}_{[i]+c[j]}(A)$ replaces row $A_{(i)}$ with $A_{(i)} + cA_{(j)}$, $c \in \mathbb{K}$.

(Type III) $\hat{R}_{u[i]}(A)$ replaces row $A_{(i)}$ with $uA_{(i)}$, u a unit in $\mathbb{K}$.

Let $\hat{C}_{[i][j]}(A)$, $\hat{C}_{[i]+c[j]}(A)$, $\hat{C}_{u[i]}(A)$ be the corresponding elementary column operations.

Remark 4.8 (**Elementary row operations as matrices**). Using 2.47 (first identity), we know that for any $n \times n$ matrix, Q, and $A \in \mathbf{M}_{n,m}(\mathbb{K})$, the rows

$$((QA)_{(1)}, \ldots, (QA)_{(n)}) = (Q_{(1)}A, \ldots, Q_{(n)}A).$$

In particular, define an $n \times n$ matrix by $Q = \hat{R}_{[i]+c[j]}(I)$ where I is the $n \times n$ identity matrix. Let $I_{(t)}$ denote row t of the identity matrix. Then, in terms of rows,

$$Q = (I_{(1)}, \ldots, I_{(i-1)}, \left[I_{(i)} + cI_{(j)}\right], I_{(i+1)}, \ldots, I_{(n)})$$
$$QA = (I_{(1)}A, \ldots, I_{(i-1)}A, \left[I_{(i)} + cI_{(j)}\right]A, I_{(i+1)}A, \ldots, I_{(n)}A)$$
$$QA = (A_{(1)}, \ldots, A_{(i-1)}, \left[A_{(i)} + cA_{(j)}\right], A_{(i+1)}, \ldots, A_{(n)})$$

Thus, $QA = \hat{R}_{[i]+c[j]}(A)$ so left multiplication of A by Q is the same as applying the elementary row operation, $\hat{R}_{[i]+c[j]}$ to A. Instead of $Q = \hat{R}_{[i]+c[j]}(I)$, we use the notation $R_{[i]+c[j]} = \hat{R}_{[i]+c[j]}(I)$ (remove the hat). Thus, $R_{[i]+c[j]} \in \mathbf{M}_{n,n}(\mathbb{K})$ is a nonsingular matrix such that $R_{[i]+c[j]}A = \hat{R}_{[i]+c[j]}(A)$.

LEMMA 4.9. *For each of the elementary row (or column) operations, $\hat{R}$ (or $\hat{C}$), there is an invertible matrix $R \in \mathbf{M}_{n,n}(\mathbb{K})$ (or $C \in \mathbf{M}_{m,m}(\mathbb{K})$) that when left-multiplied (or right-multiplied) with any matrix $A \in \mathbf{M}_{n,m}(\mathbb{K})$ results in the same matrix as $\hat{R}(A)$ (or $\hat{C}(A)$). In each case, $R = \hat{R}(I_n)$ (or $C = \hat{C}(I_m)$).*

PROOF. The argument in each case is similar to that given for $\hat{R}_{[i]+c[j]}$ in 4.8. □

An $n \times n$ elementary row matrix $R_{[i]+c[j]}$ acts by left multiplication on any $n \times m$ matrix $A \in \mathbf{M}_{n,m}(\mathbb{K})$ (for any $m \geq 0$). The corresponding elementary row operation $\hat{R}_{[i]+c[j]}$ is defined on any matrix $A \in \mathbf{M}_{n,m}(\mathbb{K})$ where $m \geq 0$ and $n \geq \max(i, j)$. This difference in natural domains between the functions $\hat{R}$ and the matrices R (by left multiplication) needs to be kept in mind in some discussions.

REMARK 4.10 (**IDENTITIES FOR ELEMENTARY ROW MATRICES**). Check the following for $n = 3$ - the case for $n \times n$ matrices is the same idea.

(Type I) $R^{-1}_{[i][j]} = R_{[i][j]},\ \ R^{T}_{[i][j]} = R_{[i][j]},\ \ \det(R_{[i][j]}) = -1$

(Type II) $R^{-1}_{[i]+c[j]} = R_{[i]-c[j]},\ \ R^{T}_{[i]+c[j]} = R_{[j]+c[i]},\ \ \det(R_{[i]+c[j]}) = 1$

(Type III) $R^{-1}_{u[i]} = R_{u^{-1}[i]},\ \ R^{T}_{u[i]} = R_{u[i]},\ \ \det(R_{u[i]}) = u$

For example, $R_{[2]+[3]} = \begin{pmatrix} 1 & 0 & 0 \\ 0 & 1 & 1 \\ 0 & 0 & 1 \end{pmatrix}$ and $R^{-1}_{[2]+[3]} = R_{[2]-[3]} = \begin{pmatrix} 1 & 0 & 0 \\ 0 & 1 & -1 \\ 0 & 0 & 1 \end{pmatrix}$.

REMARK 4.11 (**EUCLIDEAN ALGORITHM AND GREATEST COMMON DIVISORS**). We recall the Euclidean algorithm for computing $r_k = \gcd(r_0, r_1)$ where r_0 and r_1 are nonzero elements of $\mathbb{Z}$. The same algorithm works for any Euclidean domain, in particular for $\mathbb{K}$. The algorithm is usually described by a layout representing successive divisions. The layout for $\mathbb{K} = \mathbb{F}$ is trivial: $r_0 = q_1 r_1$ where $q_1 = r_0 r_1^{-1}$. Here is the general pattern:

$$\begin{array}{rcl} r_0 & = & q_1 r_1 + r_2 \\ r_1 & = & q_2 r_2 + r_3 \\ r_2 & = & q_3 r_3 + r_4 \\ \vdots & \vdots & \vdots \\ r_{k-3} & = & q_{k-2} r_{k-2} + r_{k-1} \\ r_{k-2} & = & q_{k-1} r_{k-1} + r_k \\ r_{k-1} & = & q_k r_k + 0 \end{array}$$

If $r_2 \neq 0$, the remainders, $r_2, \ldots, r_k$, have strictly decreasing valuations and thus must terminate with zero – in this case, $r_{k+1} = 0$. The last nonzero remainder, r_k in this case, is the $\gcd(r_0, r_1)$. In fact, the set of all divisors of r_k satisfies: $\{x : x \mid r_k\} = \{x : x \mid r_0\} \cap \{x : x \mid r_1\}$. This fact is easily seen (or proved by induction) from the layout above.

REMARK 4.12 (**GREATEST COMMON DIVISOR AS LINEAR COMBINATION**). Referring to Remark 4.11, the second to the last identity in the successive division layout, $r_{k-2} = q_{k-1}r_{k-1} + r_k$ can be solved for r_k to get $r_k = r_{k-2} - q_{k-1}r_{k-1} = s_{k-2}r_{k-2} + t_{k-1}r_{k-1}$ (this defines s_{k-2} and t_{k-1}). Using $r_{k-3} - q_{k-2}r_{k-2} = r_{k-1}$ to eliminate r_{k-1} gives $r_k = s_{k-3}r_{k-3} + t_{k-2}r_{k-2}$. Repeating this process (or using induction) gives $r_k = s_0r_0 + t_1r_1$. The standard theorem from basic algebra is that if a and b are nonzero elements of a Euclidean domain $\mathbb{K}$ and $d = \gcd(a, b)$ then there exists $s, t \in \mathbb{K}$ such that

$$d = sa + tb.$$

This theorem is easily proved without using the Euclidean algorithm by using the fact that the Euclidean domain $\mathbb{K}$ is also a principal ideal domain (1.20).

REMARK 4.13 (**MATRIX VERSIONS OF EUCLIDEAN ALGORITHM**). The sequence of remainders displayed in Remark 4.11 can be represented by a sequence of matrix multiplications as follows:

$$(4.14)\qquad \begin{pmatrix} r_t \\ r_{t+1} \end{pmatrix} = \begin{pmatrix} 0 & 1 \\ 1 & -q_t \end{pmatrix} \begin{pmatrix} r_{t-1} \\ r_t \end{pmatrix} \quad t = 1, \ldots, k \ \ (r_{k+1} = 0).$$

The matrix of 4.14 is a product of type I and II elementary row matrices (4.7):

$$(4.15)\qquad R_{[1][2]}R_{[1]-q_t[2]} = \hat{R}_{[1][2]}\hat{R}_{[1]-q_t[2]}I_2 = \begin{pmatrix} 0 & 1 \\ 1 & -q_t \end{pmatrix}.$$

Interpreting the order of the products as in Remark 4.20, let

$$(4.16)\qquad Q_2 = \prod_{t=1}^{k} R_{[1][2]}R_{[1]-q_t[2]} = \begin{pmatrix} r_{11} & r_{12} \\ r_{21} & r_{22} \end{pmatrix}.$$

From the Euclidean algorithm, 4.11, we have

$$(4.17)\qquad \begin{pmatrix} d \\ 0 \end{pmatrix} = \begin{pmatrix} r_{11} & r_{12} \\ r_{21} & r_{22} \end{pmatrix} \begin{pmatrix} a \\ b \end{pmatrix}.$$

An alternative point of view follows from 4.12. If a and b are nonzero elements of a Euclidean domain $\mathbb{K}$ and $d = \gcd(a, b)$ then there exists $s, t \in \mathbb{K}$ such that $sa + tb = d$ and hence

$$(4.18)\qquad \begin{pmatrix} d \\ 0 \end{pmatrix} = \begin{pmatrix} s & t \\ \frac{-b}{d} & \frac{+a}{d} \end{pmatrix} \begin{pmatrix} a \\ b \end{pmatrix}.$$

Note that $\det \begin{pmatrix} s & t \\ \frac{-b}{d} & \frac{+a}{d} \end{pmatrix} = 1$ so this matrix is a unit in $\mathbf{M}_{2,2}(\mathbb{K})$.

If we take

$$\begin{pmatrix} a \\ b \end{pmatrix} = \begin{pmatrix} 18 \\ 12 \end{pmatrix} \quad s = +1, \quad t = -1, \quad d = 6, \quad \begin{pmatrix} s & t \\ \frac{-b}{d} & \frac{+a}{d} \end{pmatrix} = \begin{pmatrix} +1 & -1 \\ -2 & +3 \end{pmatrix}$$

then

$$\begin{pmatrix} +1 & -1 \\ -2 & +3 \end{pmatrix} \begin{pmatrix} 18 \\ 12 \end{pmatrix} = \begin{pmatrix} 6 \\ 0 \end{pmatrix} = \begin{pmatrix} d \\ 0 \end{pmatrix}. \tag{4.19}$$

Remark 4.20 (**Examples of matrix versions**). Take $\begin{pmatrix} a \\ b \end{pmatrix} = \begin{pmatrix} 18 \\ 12 \end{pmatrix}$. The Euclidean algorithm has two steps: $18 = 1 \cdot 12 + 6$ ($q_1 = 1$) and $12 = 2 \cdot 6$ ($q_2 = 2$). Thus, 4.15 and 4.16 become

$$Q_2 = \begin{pmatrix} 0 & 1 \\ 1 & -2 \end{pmatrix} \begin{pmatrix} 0 & 1 \\ 1 & -1 \end{pmatrix} = \begin{pmatrix} +1 & -1 \\ -2 & +3 \end{pmatrix} = \begin{pmatrix} r_{11} & r_{12} \\ r_{21} & r_{22} \end{pmatrix}. \tag{4.21}$$

Thus, 4.17 becomes

$$\begin{pmatrix} 6 \\ 0 \end{pmatrix} = \begin{pmatrix} +1 & -1 \\ -2 & +3 \end{pmatrix} \begin{pmatrix} 18 \\ 12 \end{pmatrix}. \tag{4.22}$$

To see how these 2×2 matrices are used in general, recall the notation of 3.25 and take $Q_4 = Q_2 \oplus_X^X I_2$ to be the general direct sum corresponding to $X = Y = \{2, 4\}$. Let A be a 4×6 matrix as shown:

$$Q_4 = \begin{pmatrix} 1 & 0 & 0 & 0 \\ 0 & \underline{+1} & 0 & \underline{-1} \\ 0 & 0 & 1 & 0 \\ 0 & \underline{-2} & 0 & \underline{+3} \end{pmatrix} \quad A = \begin{pmatrix} +1 & -1 & 0 & 0 & 3 & 4 \\ +2 & 0 & \underline{18} & -1 & 2 & 6 \\ 0 & +1 & 2 & 2 & 5 & 3 \\ -5 & -4 & \underline{12} & 5 & 2 & 8 \end{pmatrix}. \tag{4.23}$$

Using 3.49, we see that Q_4 is a unit matrix in $\mathbf{M}_{4,4}(\mathbb{Z})$:

$$\det(Q_4) = (-1)^{2\sum X} \det(Q_4[X \mid X]) \det(Q_4(X \mid X)) = \det(Q_2)\det(I_2) = +1.$$

Thus, Q_4 is a unit in $\mathbf{M}_{4,4}(\mathbb{K})$. Since Q_2 is a product of elementary row operations (or matrices) so is Q_4. In fact, only type I and II matrices are needed. Consider Q_4A.

$$Q_4A = \begin{pmatrix} +1 & -1 & 0 & 0 & 3 & 4 \\ 7 & 4 & \underline{6} & -6 & 0 & -2 \\ 0 & +1 & 2 & 2 & 5 & 3 \\ -19 & -12 & \underline{0} & 17 & 2 & 12 \end{pmatrix}. \tag{4.24}$$

Note how Q_4 transforms the underlined entries in A shown in 4.23 to those shown in 4.24 and compare these transformations with 4.22. Note also, that setting $A' = Q_4A$, we have $A'(X \mid \underline{m}] = A(X \mid \underline{m}]$ where $X = \{2, 4\}$ and $m = 6$. This discussion leads to the following lemma.

Lemma 4.25. *Let $A \in \mathbf{M}_{n,m}(\mathbb{K})$, let $(A(i_1, j), \ldots, A(i_k, j))$, $i_1 < \cdots < i_k$, be k nonzero entries in column j of A, and let $s \in \{i_1, \ldots, i_k\}$ be specified. There exists a unit $Q \in \mathbf{M}_{n,n}(\mathbb{K})$ such that $QA = A'$ satisfies*

$$(A'(i_1, j), \ldots, A'(s, j), \ldots, A'(i_k, j)) = (0, \ldots, d, \ldots, 0) \tag{4.26}$$

$$(4.27) \qquad d = \gcd(A(i_1, j), A(i_2, j), \ldots, A(i_k, j))$$

$$(4.28) \qquad A'(i_1, \ldots, i_k \mid \underline{m}] = A(i_1, \ldots, i_k \mid \underline{m}] \text{ (notation 2.52).}$$

Furthermore, Q can be chosen to be a product of type I and II row operations.

Proof. The lemma is a restatement of ideas discussed in Remarks 4.13 and 4.20. It suffices to consider $s = i_1$ since repositioning d can be done by one elementary (type I) row operation. The proof is by induction on k. The case $k = 2$ is discussed in 4.13. Assume there is a product of elementary row operations $\tilde{Q}$ such that $\tilde{Q}A = \tilde{A}$ satisfies

$$\begin{aligned}
&(\tilde{A}(i_2, j), \ldots, \tilde{A}(i_k, j)) = (\tilde{d}, 0, \ldots, 0),\\
&\tilde{d} = \gcd(A(i_2, j), \ldots, A(i_k, j)), \text{ and}\\
&\tilde{A}(i_2, \ldots, i_k \mid \underline{m}] = A(i_2, \ldots, i_k \mid \underline{m}].
\end{aligned}$$

Thus, $(\tilde{A}(i_1, j), \tilde{A}(i_2, j)) = (A(i_1, j), \tilde{d})$ is the case $k = 2$. With Q_2 as in Remark 4.20, let $Q_n = Q_2 \oplus_X^X I_{n-2}$ where $X = \{i_1, i_2\}$. $Q = Q_n\tilde{Q}$ is the required unit Q such that $QA = A'$ has the properties 4.26, 4.27, and 4.28 (with $s = i_1$). □

Hermite form, canonical forms and uniqueness

Let $A \in \mathbf{M}_{n,m}(\mathbb{K})$ (1.30). We are interested in characterizing certain "nicely structured" matrices H which are left-unit equivalent to A (4.2). In particular, we study those with the general structure shown in Figure 4.29. Such matrices are called *row echelon forms* or *Hermite forms.*

(4.29) **Figure : Hermite form**

$$H = \left[\begin{array}{ccccccccccccc}
0 & \cdots & 0 & h_{1j_1} & X & \cdots & X & X & X & \cdots & X & X & X \cdots X\\
0 & \cdots & 0 & 0 & 0 & \cdots & 0 & h_{2j_2} & X & \cdots & X & X & X \cdots X\\
\vdots & \vdots & \vdots & \vdots & \vdots & & \vdots & \vdots & \vdots & & \vdots & \vdots & \vdots \quad \vdots\\
0 & \cdots & 0 & 0 & 0 & \cdots & 0 & 0 & 0 & \cdots & 0 & h_{rj_r} & X \cdots X\\
\hline
\multicolumn{13}{c}{\mathbf{0}}
\end{array}\right]$$

Definition 4.30 (**Row Hermite or row echelon form**). A matrix $H \in \mathbf{M}_{n,m}(\mathbb{K})$ is in *row Hermite form* (or *row echelon form*) if it is the zero matrix, Θ_{nm}, or it is nonzero and looks like the matrix in Figure 4.29. Specifically, for a nonzero H the following hold:

1 For some $1 \leq r \leq n$ the first r rows are nonzero; the rest are zero.

2 In each nonzero row i the first or *primary* nonzero row entry is h_{ij_i}.

3 The *primary column indices* are $j_1 < j_2 \cdots < j_r$.

The number r of nonzero rows is the rank $\rho(H)$ of H which is also the dimension of the vector space spanned by the rows or columns of H over the quotient field of $\mathbb{K}$. Note that $\det H[1,\ldots,r \mid j_1,\ldots,j_r] \neq 0$, and any $k \times k$ sub-determinant of H with $k > r$ has determinant zero. Thus r is the rank of H in the sense of 3.119 and also the rank of any matrix A left-unit equivalent to H (3.122).

In most discussions it will be clear if we are talking about "row" Hermite form or the alternative "column" Hermite form. We will prove that for $n, m \geq 1$, any matrix in $\mathbf{M}_{n,m}(\mathbb{K})$ is left-unit equivalent to a matrix in Hermite form. The general proof for matrices in $\mathbf{M}_{n,m}(\mathbb{K})$ is by induction on m, having established the case $m = 1$ for $n \geq 1$.

THEOREM 4.31 (**HERMITE FORM**). *Let $A \in \mathbf{M}_{n,m}(\mathbb{K})$ (1.30). There exists a unit $Q \in \mathbf{M}_{n,n}(\mathbb{K})$ such that $QA = H$ where H is a Hermite (row echelon) form (4.30). Q is a product of type I and II elementary row matrices.*

PROOF. Suppose $m = 1$ and $n \geq 1$. If $A = \Theta_{n,1}$ then it is a Hermite form by definition. Suppose $A \neq \Theta_{n,1}$ contains k nonzero entries. Apply 4.25 with $j = 1$, $s = i_1$ where $(A(i_1,1),\ldots,A(i_k,1))$ contains all of the nonzero entries in column 1. Thus we obtain a unit Q' such that $A' = Q'A$ has $(A'(i_1,1),\ldots,\ldots,A'(i_k,1)) = (d,0,\ldots,0)$. Apply $\hat{R}_{[1][i_1]}I_n$ (if $1 < i_1$) to obtain the unit matrix $Q \in \mathbf{M}_{n,n}(\mathbb{K})$ such that $QA = H$ is a nonzero $n \times 1$ matrix in Hermite form. Q is a product of type I and II elementary row matrices.

By induction on m, assume there is unit matrix P, a product of type I and II elementary row matrices, such that PB is a Hermite form for any matrix $B \in \mathbf{M}_{n,m-1}(\mathbb{K})$, $n \geq 1$. Now suppose $A \in \mathbf{M}_{n,m}(\mathbb{K})$ where $m > 1$. Either (1) $A^{(1)} = \Theta_{n1}$ or (2) there is unit matrix $\tilde{Q}$, a product of type I and II elementary row matrices, such that $\tilde{Q}A = \tilde{A}$ has $\tilde{A}^{(1)} = (d,0,\ldots,0)$, $d \neq 0$.

In case (1), by the induction hypothesis, there is a unit $Q \in \mathbf{M}_{n,n}(\mathbb{K})$ such that $QA[1,\ldots,n \mid 2,\ldots,m]$ is a Hermite form and thus QA is a Hermite form. Q is a product of type I and II elementary row matrices.

In case (2), there is a unit $\tilde{P} \in \mathbf{M}_{n-1,n-1}(\mathbb{K})$ such that $\tilde{P}\tilde{A}[2,\ldots,n \mid 2,\ldots,m]$ is a Hermite form. Thus, $Q = ((1) \oplus \tilde{P})\tilde{Q}$ is such that QA is a Hermite form. By the induction hypothesis, $\tilde{P}$ (hence $(1) \oplus \tilde{P}$) and $\tilde{Q}$ are products of type I and II elementary row operations and hence so is Q.

□

REMARK 4.32 (**HERMITE FORM OF A UNIT MATRIX**). Suppose in 4.31 $m = n$ and the matrix $A \in \mathbf{M}_{n,n}(\mathbb{K})$ is a unit. Thus any Hermite form, $H = Q'A$, must be a unit and upper triangular: $H(i,j) = 0$ if $i > j$. Thus all diagonal elements of H are units. By using additional type I elementary row operations there is a unit P such that $PH = H'$ is a diagonal matrix with units along the diagonal. By using additional type III elementary row operations, there is a unit P' such that $P'H' = I_n$. Thus, for any unit matrix A there is a product of elementary row operations of type I and II that reduces A to a diagonal matrix and a product of type I, II, and III elementary row operations that reduces A to the identity. In particular, $A \in \mathbf{M}_{n,n}(\mathbb{K})$ is a unit if and only if it is a product of elementary row operations (or matrices). The statement "row operations" can be replaced by "column operations."

COROLLARY 4.33. *$A \in \mathbf{M}_{n,n}(\mathbb{K})$ is a unit if and only if it is a product of elementary row operations (or matrices).*

PROOF. See the discussion of 4.32. Note that, in general, type I, II, and III elementary row operations are required. □

(4.34) Uniqueness results for Hermite forms

We now discuss additional structural conditions on Hermite forms that make them unique (or *canonical*). Note that if $A \in \mathbf{M}_{n,m}(\mathbb{K})$ and $\tilde{P}, \tilde{Q} \in \mathbf{M}_{n,n}(\mathbb{K})$ are units such that $\tilde{P}A = K$ and $\tilde{Q}A = H$ where K and H are Hermite forms (4.30). Then, $PH = K$ where $P = \tilde{P}\tilde{Q}^{-1}$.

LEMMA 4.35 (**LEFT-UNIT EQUIVALENCE OF HERMITE FORMS**). *Suppose the Hermite forms $H, K \in \mathbf{M}_{n,m}(\mathbb{K})$ are left-unit equivalent: $PH = K$. Then the primary column indices of H and K are the same.*

PROOF. We use Lemma 4.3 which states that linear relations among column vectors are preserved under left-unit equivalence. Assume first that $\mathbb{K}$ is field. Let $(j_1, \ldots, j_r)$ and $(i_1, \ldots, i_r)$ be the primary column indices for H and K respectively. If $(j_1, \ldots, j_r)$ does not equal $(i_1, \ldots, i_r)$, then, without loss of generality, let t be the first index such that $j_t < i_t$ (possibly $t = 1$.) Then columns $H^{(j_1)}, \ldots, H^{(j_t)}$ are linearly independent but columns $K^{(j_1)}, \ldots, K^{(j_t)}$ are not linearly independent. Thus, $(j_1, \ldots, j_r) = (i_1, \ldots, i_r)$. If $\mathbb{K}$ is not a field, apply the same argument to the quotient field of $\mathbb{K}$, noting that $PH = K$ is a valid identity in the quotient field. □

Recall Definition 1.33 and the discussion following it.

DEFINITION 4.36 (**CANONICAL SDR FOR ASSOCIATES**). For $s, t \in \mathbb{K}$, define an equivalence relation (1.35) by $s \sim t$ if $us = t$ for some unit $u \in \mathbb{K}$. If $s \sim t$ then s and t are *associates* in $\mathbb{K}$. Otherwise, s and t are *nonassociates.* If $\mathbb{K} = \mathbb{Z}$ define the SDR for associates to be the set $\{n \mid n \geq 0\}$. If $\mathbb{K} = \mathbb{F}[x]$ the SDR for associates is the zero polynomial and all monic polynomials (i.e., $a_k x^k + \cdots + a_0$ with $a_k = 1$, $1x^0 = 1$). If $\mathbb{K} = \mathbb{F}$ is a field the SDR for associates is $\{0, 1\}$. These SDRs are also called *complete systems of nonassociates.*

Definition 4.37 (**Canonical SDR for residues**). Given $0 \neq m \in \mathbb{K}$, define an equivalence relation on $\mathbb{K}$ by $s \sim t$ if $m \mid (s-t)$ (m divides $s-t$). If $s \sim t$ then s and t are equivalent modulo m. If $\mathbb{K} = \mathbb{Z}$ the *(canonical)* SDR for residues modulo $m > 0$ is $\{0, 1, \ldots, m-1\}$. If $\mathbb{K} = \mathbb{F}[x]$ the SDR for residues modulo m is $\{0\} \cup \{p(x) \mid \deg(p(x)) < \deg(m(x))\}$. If $\mathbb{K} = \mathbb{F}$ the SDR for residues modulo m is $\{0\}$. These SDRs are also called *complete systems of residues modulo m.*

Definition 4.38 (**Hermite canonical form – row version**). Let $H \in \mathbf{M}_{n,m}(\mathbb{K})$ be a row Hermite form (4.30). Suppose the primary row entries, h_{tj_t}, $1 \leq t \leq r$, are elements of the canonical SDR for associates for $\mathbb{K}$ (4.36) and the h_{ij_t}, $i < t$, are elements of the canonical SDR for residues modulo h_{tj_t}, $1 \leq t \leq r$. Then the Hermite form H is called a Hermite *canonical* form.

Remark 4.39 (**Computing Hermite canonical form**). Let $Q \in \mathbf{M}_{n,n}(\mathbb{K})$ be a unit and suppose that $QA = H$ where H is a Hermite form (4.30). By using elementary row operations of the form $\hat{R}_{u[t]}$, u a unit, we can transform the primary row entries, h_{tj_t}, $1 \leq t \leq r$, such that they are elements of the SDR for associates for $\mathbb{K}$ (4.36). Next, by using elementary row operations of the form $\hat{R}_{[i]-q_{ij_t}[t]}$, $1 \leq i < t$ $(2 \leq t \leq r)$, on the resulting H, we can arrange that h_{i,j_t} are in the SDR for residues modulo h_{tj_t} (4.37). The SDR for associates phase is done first, then the SDR for residues phase. The residues are computed left to right. Figure 4.40 shows this computation where the residues already computed are indicated by h'. The last residues, corresponding to dividing by h_{rj_r} (by applying $\hat{R}_{[i]-q_{ij_r}[r]}$ as needed), are yet to be computed.

(4.40) **Figure : Compute residue phase**

$$\begin{bmatrix} 0 & \cdots & h_{1j_1} & *** & h'_{1j_2} & *** & h'_{1j_{r-1}} & *** & h_{1j_r} & *** & * \\ 0 & \cdots & 0 & \cdots & h_{2j_2} & *** & h'_{2j_{r-1}} & *** & h_{2j_r} & *** & * \\ \vdots & \vdots & \vdots & \vdots & \vdots & \vdots & \vdots & \vdots & \vdots & \vdots & \vdots \\ 0 & \cdots & 0 & \cdots & 0 & \cdots & h_{(r-1)j_{r-1}} & *** & h_{(r-1)j_r} & *** & * \\ 0 & \cdots & 0 & \cdots & 0 & \cdots & 0 & \cdots & h_{rj_r} & *** & h_{rm} \end{bmatrix}$$

Remark 4.41 (**Examples of Hermite canonical forms**). We give three examples. In these examples, we omit the initial zero columns and the terminal zero rows. The initial, nonzero, elements of the rows (the h_{ij_i}) are referred to as the primary row entries (sometimes called "pivots"). For the definitions of the SDR for associates and the SDR for residues see 4.36, 4.37.

In our first example, 4.42, let $\mathbb{K} = \mathbb{F}$ be a field. In the Hermite canonical form for a field, the primary row entries are all 1. The elements above the these pivots are all zero since the SDR for residues is $\{0\}$ in a field.

$$\left[\begin{array}{ccccccccc} 1 & *** & 0 & *** & 0 & *** & 0 & *** & * \\ 0 & \cdots & 1 & *** & 0 & *** & 0 & *** & * \\ \vdots & \vdots & \vdots & \vdots & \vdots & \vdots & \vdots & \vdots & \vdots \\ 0 & \cdots & 0 & \cdots & 1 & *** & 0 & *** & * \\ 0 & \cdots & 0 & \cdots & 0 & \cdots & 1 & *** & * \end{array}\right] \quad \mathbb{K} = \mathbb{F} \tag{4.42}$$

In the second example, 4.43, let $\mathbb{K} = \mathbb{Z}$ be the integers. In the Hermite canonical form for $\mathbb{Z}$, the pivots are all nonzero and belong to the SDR for associates: $\{n \mid n \geq 0\}$. The elements above the pivots h_{tj_t} are all in the SDR for residues modulo h_{tj_t}: $\{0, 1, \ldots, h_{tj_t} - 1\}$.

$$\left[\begin{array}{ccccccccc} 3 & *** & 5 & *** & 1 & *** & 6 & *** & * \\ 0 & \cdots & 6 & *** & 3 & *** & 0 & *** & * \\ \vdots & \vdots & \vdots & \vdots & \vdots & \vdots & \vdots & \vdots & \vdots \\ 0 & \cdots & 0 & \cdots & 5 & *** & 8 & *** & * \\ 0 & \cdots & 0 & \cdots & 0 & \cdots & 9 & *** & * \end{array}\right] \quad \mathbb{K} = \mathbb{Z} \tag{4.43}$$

In the third example, 4.44, let $\mathbb{K} = \mathbb{F}[x]$ be the polynomials over the field $\mathbb{F}$. In the Hermite canonical form for $\mathbb{F}[x]$, the pivots are all nonzero monic polynomials. The elements above the pivots h_{tj_t} are in the SDR for residues modulo h_{tj_t}: $\{0\} \cup \{p(x) \mid \deg(p(x)) < \deg(h_{tj_t})\}$.

$$\left[\begin{array}{ccccccc} x^3+3 & *** & x^5 & *** & 2x+1 & *** & x^2+6 \\ 0 & \cdots & 0 & \cdots & x^2-1 & *** & -2x^2 \\ \vdots & \vdots & \vdots & \vdots & \vdots & \vdots & \vdots \\ 0 & \cdots & 0 & \cdots & 0 & \cdots & x^3-8 \end{array}\right] \quad \mathbb{K} = \mathbb{F}[x] \tag{4.44}$$

Recall Lemma 4.35 which showed that if H and K are Hermite forms and $PH = K$ then the primary column indices of H and K are the same. If H and K are Hermite *canonical* forms then the result is much stronger. In what follows, $P \in \mathbf{M}_{n,n}(\mathbb{K})$ is a unit and $H, K \in \mathbf{M}_{n,m}(\mathbb{K})$ are Hermite canonical forms. The primary column indices of H and K are $j_1 < j_2 < \cdots < j_r$ where r is the number of non zero rows in H and K. If $\mathbb{K}$ is a field, r is the rank (row rank equals column rank) of H and K.

Remark 4.45 (**Uniqueness of Hermite canonical form with $r = 1$**). Assume $PH = K$ where H and K are $n \times m$ Hermite canonical forms and P is a unit in $\mathbf{M}_{n,n}(\mathbb{K})$ (which is equivalent to $\det(P)$ a unit in $\mathbb{K}$). If $K = \Theta_{nm}$

then $PH = K = \Theta_{nm}$ implies $H = \Theta_{nm}$ for any unit $P \in \mathbf{M}_{n,n}(\mathbb{K})$. Thus, we consider the case where $K \neq \Theta$ and take $r = 1$, $P = (p_{ij})$ and $PH = K$ where

$$(4.46)\qquad K = \begin{bmatrix} 0 & \cdots & 0 & k_{1j_1} & * & \cdots & * & k_{1m} \\ 0 & \cdots & 0 & 0 & 0 & \cdots & 0 & 0 \\ \vdots & \cdots & \vdots & \vdots & \vdots & \cdots & \vdots & \vdots \\ 0 & \cdots & 0 & 0 & 0 & \cdots & 0 & 0 \end{bmatrix}.$$

Note that

$$(4.47)\qquad (PH)^{(j_1)} = PH^{(j_1)} = \begin{bmatrix} p_{11}h_{1j_1} \\ p_{21}h_{1j_1} \\ \vdots \\ p_{n1}h_{1j_1} \end{bmatrix} = \begin{bmatrix} k_{1j_1} \\ 0 \\ \vdots \\ 0 \end{bmatrix}.$$

Identity 4.47 implies two important facts:

(1) Since $p_{s1}h_{1j_1} = 0$, $s = 2, \ldots, n$, and $h_{1j_1} \neq 0$, we have $p_{s1} = 0$, $s = 2, \ldots, n$. Thus, (1) implies that $\det(P) = p_{11}\det(P(1 \mid 1))$. Since $\det(P)$ is a unit in $\mathbb{K}$, both p_{11} and $\det(P(1 \mid 1))$ are units (1.16).

(2) Since $p_{11}h_{1j_1} = k_{1j_1}$ and p_{11} is a unit, the fact that h_{1j_1} and k_{1j_1} belong to the same SDR for associates for $\mathbb{K}$ implies that $p_{11} = 1$.

Hence, P has the following structure:

$$(4.48)\qquad P = \begin{bmatrix} 1 & p_{12} & \cdots & p_{1n} \\ 0 & \cdot & \cdots & \cdot \\ \vdots & \vdots & P(1|1) & \vdots \\ 0 & \cdot & \cdots & \cdot \end{bmatrix} = \begin{bmatrix} I_1 & P[1|1) \\ \Theta_{n-1,1} & P(1|1) \end{bmatrix}.$$

where $P(1|1) \in \mathbf{M}_{n-1,n-1}(\mathbb{K})$ is a unit and $P[1|1) \in \mathbf{M}_{1,n-1}(\mathbb{K})$ is arbitrary. Thus, since H is a rank one Hermite canonical form,

$$(4.49)\qquad PH = \begin{bmatrix} 0 & \cdots & 0 & h_{1j_1} & * & \cdots & * & h_{1m} \\ 0 & \cdots & 0 & 0 & 0 & \cdots & 0 & 0 \\ \vdots & \cdots & \vdots & \vdots & \vdots & \cdots & \vdots & \vdots \\ 0 & \cdots & 0 & 0 & 0 & \cdots & 0 & 0 \end{bmatrix} = K.$$

implies that $h_{1t} = k_{1t}$ for $t = j_1, \ldots, m$ and thus $H = K$.

Remark 4.50 (**Uniqueness of Hermite canonical form $2 \times m$ case**). Assume $PH = K$ where $P \in \mathbf{M}_{2,2}(\mathbb{K})$ is a unit and $H, K \in \mathbf{M}_{2,m}(\mathbb{K})$ are $2 \times m$ Hermite canonical forms. Thus, in this example, $n = r = 2$, $P = \begin{bmatrix} p_{11} & p_{12} \\ p_{21} & p_{22} \end{bmatrix}$ and $PH = K$ where

$$(4.51)\qquad H = \begin{bmatrix} 0 & \cdots & 0 & h_{1j_1} & * & \cdots & * & h_{1j_2} & * & \cdots & * \\ 0 & \cdots & 0 & 0 & 0 & \cdots & 0 & h_{2j_2} & * & \cdots & * \end{bmatrix}.$$

Note that the first primary column $K^{(j_1)} = (PH)^{(j_1)} = P(H^{(j_1)})$. Thus,

$$\begin{bmatrix} k_{1j_1} \\ 0 \end{bmatrix} = \begin{bmatrix} p_{11}h_{1j_1} \\ p_{21}h_{1j_1} \end{bmatrix} \tag{4.52}$$

which implies (since $h_{1j_1} \neq 0$) that $p_{21} = 0$. Since $\det(P) = p_{11}p_{22}$ is a unit, p_{11} and p_{22} are units in $\mathbb{K}$. Since k_{1j_1} and h_{1j_1} are assumed to be from the same SDR for associates of $\mathbb{K}$ (4.36), we have $p_{11} = 1$ and $k_{1j_1} = h_{1j_1}$. At this point, $P = \begin{bmatrix} 1 & p_{12} \\ 0 & p_{22} \end{bmatrix}$ and

$$PH = \begin{bmatrix} 0 & \cdots & 0 & k_{1j_1} & * & \cdots & * & h_{1j_2} + p_{12}h_{2j_2} & * & \cdots & * \\ 0 & \cdots & 0 & 0 & 0 & \cdots & 0 & p_{22}h_{2j_2} & * & \cdots & * \end{bmatrix}. \tag{4.53}$$

From $PH = K$ we get that $p_{22}h_{2j_2} = k_{2j_2}$ and since $h_{2j_2}$and k_{2j_2} belong the same SDR for associates, the unit $p_{22} = 1$ and $h_{2j_2} = k_{2j_2}$. Thus, we have $h_{1j_2} + p_{12}k_{2j_2} = k_{1j_2}$. But, k_{1j_2} is in the SDR for residues modulo k_{2j_2} (4.37). Thus, $p_{12} = 0$, $P = I_2$ and $H = K$.

Remarks 4.45 and 4.50 illustrate all of the ideas needed for the general proof. We use the standard submatrix notation 2.52.

Theorem 4.54 (Uniqueness of Hermite canonical form). *If $PH = K$ where $P \in \mathbf{M}_{n,n}(\mathbb{K})$ is a unit and $H, K \in \mathbf{M}_{n,m}(\mathbb{K})$ are Hermite canonical forms then $H = K$ and P is of the form*

$$P = \begin{bmatrix} I_r & P[\underline{r}|\underline{r}) \\ \Theta_{n-r,r} & P(\underline{r}|\underline{r}) \end{bmatrix} \tag{4.55}$$

where r is the rank of H and K, $P(\underline{r}|\underline{r}) \in \mathbf{M}_{n-r,n-r}(\mathbb{K})$ is a unit and $P[\underline{r}|\underline{r}) \in \mathbf{M}_{r,n-r}(\mathbb{K})$.

Proof. The proof is by induction on r where $r = 0$ is trivial. The case $r = 1$ was proved in Remark 4.45. Assume $r > 1$ and the theorem is true for the case $r - 1$. We know the primary row indices, $j_1 < j_2 < \cdots < j_r$, are the same for H and K. Thus,

$$(PH)^{(j_1)} = PH^{(j_1)} = \begin{bmatrix} p_{11}h_{1j_1} \\ p_{21}h_{1j_1} \\ \vdots \\ p_{n1}h_{1j_1} \end{bmatrix} = \begin{bmatrix} k_{1j_1} \\ 0 \\ \vdots \\ 0 \end{bmatrix}. \tag{4.56}$$

Equation 4.56 implies (see 4.52 for the idea) that

(1) Since $p_{s1}h_{1j_1} = 0$, $s = 2, \ldots, n$, and $h_{1j_1} \neq 0$, we have $p_{s1} = 0$, $s > 1$. Thus $p_{11}\det(P(1\,|\,1)) = \det(P)$. This implies that p_{11} and $\det(P(1\,|\,1))$ are units (1.23). We also have that

(2) $p_{11}h_{1j_1} = k_{1j_1}$, and since h_{1j_1} and k_{1j_1} belong to the same SDR for associates (4.36), $p_{11} = 1$.

Thus, (1) and (2) imply that $\det(P) = \det(P(1 \mid 1))$ (2.55). Hence, $P(1|1)$ is a unit in $\mathbf{M}_{n-1,n-1}(\mathbb{K})$ (3.84). P has the following structure:

$$(4.57) \qquad P = \begin{bmatrix} 1 & p_{12} & \cdots & p_{1n} \\ 0 & \cdot & \cdots & \cdot \\ \vdots & \vdots & P(1|1) & \vdots \\ 0 & \cdot & \cdots & \cdot \end{bmatrix} = \begin{bmatrix} I_1 & P[1|1) \\ \Theta_{n-1,1} & P(1|1) \end{bmatrix},$$

where $P(1|1) \in \mathbf{M}_{n-1,n-1}(\mathbb{K})$ is a unit and $P[1|1) \in \mathbf{M}_{1,n-1}(\mathbb{K})$. Thus we have

$$(4.58) \qquad PH = \begin{bmatrix} H[1|1,\ldots,j_1] & (PH)[1|1,\ldots,j_1) \\ \Theta_{n-1,j_1} & P(1|1)H(1|1,\ldots,j_1) \end{bmatrix} = K.$$

Equation 4.58 implies that $H[1|1,\ldots,j_1] = [0 \cdots 0 \;\; k_{j_1}]$, $(PH)[1|1,\ldots,j_1) = K[1|1,\ldots,j_1)$ and

$$(4.59) \qquad P(1|1)H(1|1,\ldots,j_1) = K(1|1,\ldots,j_1).$$

By the induction hypothesis, 4.59 gives $H(1|1,\ldots,j_1) = K(1|1,\ldots,j_1)$ since $P(1|1) \in \mathbf{M}_{n-1,n-1}(\mathbb{K})$ is a unit and $H(1|1,\ldots,j_1)$ and $K(1|1,\ldots,j_1)$ are Hermite canonical forms in $\mathbf{M}_{n-1,m-j_1}(\mathbb{K})$.

We claim that $(PH)[1|1,\ldots,j_1) = K[1|1,\ldots,j_1)$ implies that $p_{12} = \cdots = p_{1r} = 0$. Otherwise, let t be the first integer such that $p_{1t} \neq 0$, $2 \leq t \leq r$. This implies (see 4.53 for basic idea)

$$k_{1t} = PH(1,t) = p_{11}h_{1t} + p_{1t}k_{tj_t} = h_{1t} + p_{1t}k_{tj_t} \quad (\text{as } p_{11} = 1)$$

which contradicts the fact that k_{1t} and h_{1t} are in the canonical SDR for residues modulo $k_{tj_t} = h_{tj_t}$.

Finally, note that 4.59 implies that by the induction hypothesis, $P(1|1)$ has the structure of 4.55 with n replaced by $n-1$, r replaced by $r-1$ and P replaced by $P(1|1)$. This completes the proof. □

Stabilizers of GL(n, $\mathbb{K}$); column Hermite forms

Recall that the group of units of the ring $\mathbf{M}_{n,n}(\mathbb{K})$ is denoted by $\mathrm{GL}(n,\mathbb{K})$ and is called the *general linear group* of $\mathbf{M}_{n,n}(\mathbb{K})$ (4.1).

DEFINITION 4.60 (**STABILITY SUBGROUPS FOR LEFT UNIT MULTIPLICATION**). The subgroup $\{Q \mid Q \in \mathrm{GL}(n,\mathbb{K}), QA = A\}$ is called the *stabilizer* or *stability* subgroup of $\mathrm{GL}(n,\mathbb{K})$ at $A \in \mathbf{M}_{n,m}(\mathbb{K})$. We denote this subgroup by $\mathrm{GL}_A(n,\mathbb{K})$.

REMARK 4.61 (**CONJUGATE STABILITY SUBGROUPS**). Note that if $QA = B$ then $\mathrm{GL}_A(n,\mathbb{K}) = Q\mathrm{GL}_B(n,\mathbb{K})Q^{-1}$. To prove this identity, note that $X \in \mathrm{GL}_B(n,\mathbb{K})$ if and only if $XB = B$ if and only if $XQA = QA$ if and only if $Q^{-1}XQA = A$ if and only if $Q^{-1}XQ \in \mathrm{GL}_A(n,\mathbb{K})$. Thus, if A and B are left unit equivalent ($QA = B$), their stability subgroups are conjugate: $\mathrm{GL}_A(n,\mathbb{K}) = Q\mathrm{GL}_B(n,\mathbb{K})Q^{-1}$. Alternatively stated, if $A \sim B$ under left unit equivalence than the stability subgroups of A and B are conjugate.

Remark 4.62 (**Stability subgroup of Hermite canonical form**). Let $H \in \mathbf{M}_{n,m}(\mathbb{K})$ be a Hermite canonical form. As a consequence of Theorem 4.54 we know that if $PH = H$ then P is of the form 4.55. Conversly, if P is of that form then $PH = H$. Thus, we have characterized $\mathrm{GL}_H(n, \mathbb{K})$. Block multiplication is as follows: $PQ =$

$$\begin{bmatrix} I_r & P[\underline{r}|\underline{r}) \\ \Theta_{n-r,r} & P(\underline{r}|\underline{r}) \end{bmatrix} \begin{bmatrix} I_r & Q[\underline{r}|\underline{r}) \\ \Theta_{n-r,r} & Q(\underline{r}|\underline{r}) \end{bmatrix} = \begin{bmatrix} I_r & Q[\underline{r}|\underline{r}) + P[\underline{r}|\underline{r})Q(\underline{r}|\underline{r}) \\ \Theta_{n-r,r} & P(\underline{r}|\underline{r})Q(\underline{r}|\underline{r}) \end{bmatrix}.$$

The inverse of P is constructed as follows:

$$P^{-1} = \begin{bmatrix} I_r & -P[\underline{r}|\underline{r})P^{-1}(\underline{r}|\underline{r}) \\ \Theta_{n-r,r} & P^{-1}(\underline{r}|\underline{r}) \end{bmatrix}.$$

We summarize with the following corollary.

Corollary 4.63 (**Characterization of stabilizer $\mathrm{GL}_A(n, \mathbb{K})$**). *Let $A \in \mathbf{M}_{n,m}(\mathbb{K})$ and let $QA = H$ where H is the Hermite canonical form of A. Assume the rank of A (and hence H) is r. The stabilizer $\mathrm{GL}_A(n, \mathbb{K}) = Q\mathrm{GL}_H(n, \mathbb{K})Q^{-1}$ where $\mathrm{GL}_H(n, \mathbb{K})$ is the set of all matrices P of the form*

$$P = \begin{bmatrix} I_r & P[\underline{r}|\underline{r}) \\ \Theta_{n-r,r} & P(\underline{r}|\underline{r}) \end{bmatrix}$$

where $P(\underline{r}|\underline{r}) \in \mathbf{M}_{n-r,n-r}(\mathbb{K})$ is an arbitrary unit and $P[\underline{r}|\underline{r}) \in \mathbf{M}_{r,n-r}(\mathbb{K})$ is an arbitrary matrix.

Proof. The proof follows from the discussion in 4.61 and 4.62.

□

(4.64) Right unit equivalence and column Hermite form

All of the results concerning left unit equivalence and row Hermite forms have direct analogs for right unit equivalence and column Hermite forms. We discuss the key results here.

Remark 4.65 (**Column Hermite form**). Suppose $\tilde{A} \in \mathbf{M}_{n,m}(\mathbb{K})$ and $\tilde{Q} \in \mathrm{GL}_n(\mathbb{K})$. Suppose $\tilde{Q}\tilde{A} = \tilde{H} \in \mathbf{M}_{n,m}(\mathbb{K})$ is a *row* Hermite form of A. Taking transposes, $\tilde{A}^T = A$, $\tilde{Q}^T = Q$, $\tilde{H}^T = H$, we have $AQ = H$ where $Q \in \mathrm{GL}_n(\mathbb{K})$. The matrix $H \in \mathbf{M}_{m,n}(\mathbb{K})$ is a *column* Hermite form of $A \in \mathbf{M}_{m,n}(\mathbb{K})$. The structure of H is shown in 4.66 (columns and rows of zeros in bold type can be repeated):

$$
(4.66)\qquad H = \begin{bmatrix}
\mathbf{0} & \mathbf{0} & \cdots & \mathbf{0} & \mathbf{0} \\
h_{j_1 1} & 0 & \cdots & 0 & \mathbf{0} \\
* & \vdots & \cdots & \vdots & \vdots \\
* & & & & \\
* & 0 & \cdots & 0 & \mathbf{0} \\
h_{j_2 1} & h_{j_2 2} & \cdots & 0 & \mathbf{0} \\
* & * & \cdots & \vdots & \vdots \\
* & * & & & \\
* & * & \cdots & \vdots & \vdots \\
* & * & & & \\
h_{j_r 1} & h_{j_r 2} & \cdots & h_{j_r r} & \mathbf{0} \\
* & * & \cdots & * & \vdots \\
* & * & & * & \\
h_{m1} & h_{m2} & \cdots & h_{mr} & \mathbf{0}
\end{bmatrix}.
$$

Referring to 4.66 , we rephrase definition 4.30:

DEFINITION 4.67 (**COLUMN HERMITE OR COLUMN ECHELON FORM**). A matrix $H \in \mathbf{M}_{m,n}(\mathbb{K})$ is in *column Hermite form* if it is the zero matrix, Θ_{mn}, or it is nonzero and looks like the matrix in Figure 4.66. Specifically, for a nonzero H the following hold:

1 For some $1 \le r \le n$ the first r columns are nonzero; the rest are zero.

2 In each nonzero column i the first nonzero or *primary* column entry is $h_{j_i i}$.

3 The *primary row indices* satisfy $j_1 < j_2 \cdots < j_r$.

The number r of nonzero columns is the rank $\rho(H)$ of H which is also the dimension of the vector space spanned by the colums or rows of H over the quotient field of $\mathbb{K}$. Note that $\det H[j_1, \ldots, j_r \mid 1, \ldots, r] \ne 0$, and any $k \times k$ sub-determinant of H with $k > r$ has determinant zero. Thus r is the rank of H in the sense of 3.119 and also the rank of any matrix A right-unit equivalent to H (3.122).

DEFINITION 4.68 (**HERMITE CANONICAL FORM – COLUMN VERSION**). Let $H \in \mathbf{M}_{m,n}(\mathbb{K})$ be a column Hermite form – column version (4.67). Suppose the primary column entries, $h_{j_t t}$, $1 \le t \le r$, are elements of the canonical SDR for associates for $\mathbb{K}$ (4.36) and the $h_{j_t i}$, $i < t$, are elements of the canonical SDR for residues modulo $h_{j_t t}$, $1 \le t \le r$. Then the Hermite form H is called a Hermite *canonical* form – column version.

REMARK 4.69 (**ROW COLUMN CANONICAL FORM UNIQUENESS**). If we reduce $A \in \mathbf{M}_{m,n}(\mathbb{K})$ to Hermite canonical form – column version, $AQ = H$ (4.38), then the primary column entries $h_{j_i, i}$ are uniquely determined. If we now reduce H to Hermite canonical form – row version, PH, (4.68) we get a matrix of the following form: $PAQ =$

$$\tag{4.70} \begin{bmatrix} d_1 & 0 & \cdots & 0 & 0 & \cdots & 0 \\ 0 & d_2 & \cdots & 0 & 0 & \ldots & 0 \\ \vdots & \vdots & \ddots & \vdots & \vdots & \vdots & \vdots \\ 0 & 0 & \cdots & d_{r-1} & 0 & \cdots & 0 \\ 0 & 0 & \cdots & 0 & d_r & \cdots & 0 \\ \mathbf{0} & \mathbf{0} & \cdots & \mathbf{0} & \mathbf{0} & \cdots & \mathbf{0} \end{bmatrix} = \left[\begin{array}{c|c} D_r & \Theta_{r,n-r} \\ \hline \Theta_{m-r,r} & \Theta_{m-r,n-r} \end{array}\right]$$

where $r = \rho(A)$, the rank of A. The matrix D_r is uniquely determined with its entries in the canonical SDR for associates in $\mathbb{K}$ (4.36). By further use of elementary row and column operations, we can put the diagonal entries of D_r in any order. Diagonalization of matrices will be discussed in section 4.

Remark 4.71 (**Row/column canonical form when $\mathbb{K} = \mathbb{F}$**). Let $A \in \mathbf{M}_{m,n}(\mathbb{K})$ where $\mathbb{K} = \mathbb{F}$ is a field have rank r. From 4.42 we know that the row Hermite canonical form has the structure shown in 4.72 (where boldface rows and columns of zeros can be repeated):

$$\tag{4.72} \begin{bmatrix} \mathbf{0} & 1 & *** & 0 & *** & 0 & *** & 0 & *** & * \\ \mathbf{0} & 0 & \cdots & 1 & *** & 0 & *** & 0 & *** & * \\ \vdots & \vdots & \vdots & \vdots & \vdots & \vdots & \vdots & \vdots & \vdots & \vdots \\ \mathbf{0} & 0 & \cdots & 0 & \cdots & 1 & *** & 0 & *** & * \\ \mathbf{0} & 0 & \cdots & 0 & \cdots & 0 & \cdots & 1 & *** & * \\ \mathbf{0} & \mathbf{0} & \cdots & \mathbf{0} & \cdots & \mathbf{0} & \cdots & \mathbf{0} & \cdots & \mathbf{0} \end{bmatrix}$$

Using elementary column operations (4.7) of type I and II (i.e., $\hat{C}_{[i][j]}$ and $\hat{C}_{[i]+c[j]}$) we can reduce 4.72 to 4.73:

$$\tag{4.73} \begin{bmatrix} 1 & 0 & \cdots & 0 & 0 & \cdots & 0 \\ 0 & 1 & \cdots & 0 & 0 & \ldots & 0 \\ \vdots & \vdots & \ddots & \vdots & \vdots & \vdots & \vdots \\ 0 & 0 & \cdots & 1 & 0 & \cdots & 0 \\ 0 & 0 & \cdots & 0 & 1 & \cdots & 0 \\ \mathbf{0} & \mathbf{0} & \cdots & \mathbf{0} & \mathbf{0} & \cdots & \mathbf{0} \end{bmatrix} = \left[\begin{array}{c|c} I_r & \Theta_{r,n-r} \\ \hline \Theta_{m-r,r} & \Theta_{m-r,n-r} \end{array}\right]$$

Finite dimensional vector spaces and Hermite forms

In this section the ring $\mathbb{K}$ is a field $\mathbb{F}$. Suppose V is a vector space over $\mathbb{F}$ with finite dimension $\dim(V)$. If $U \subset V$ is a proper subspace of V then $\dim(U) < \dim(V)$; a proper subspace of a vector space has dimension *strictly* smaller than that space. This strict decrease of dimension (rank) with proper inclusion of subspaces (submodules) is not generally true for modules over rings (see 1.28).

Let V and W be vector spaces and let $\mathbb{L}(V, W)$ denote the linear transformations from V to W. $\mathbb{L}(V, W)$ is also designated by $\mathrm{Hom}(V, W)$ (vector space homomorphisms).

Definition 4.74 (**Matrix of a pair of bases**). Let $\mathbf{v} = (v_1, \ldots v_q)$ and $\mathbf{w} = (w_1, \ldots, w_r)$ be ordered bases for V and W respectively. Suppose for each j, $1 \le j \le q$, $T(v_j) = \sum_{i=1}^{r} a_{ij} w_i$. The matrix $A = (a_{ij})$ is called the matrix of T with respect to the base pair $(\mathbf{v}, \mathbf{w})$. We write $[T]_{\mathbf{v}}^{\mathbf{w}}$ for A.

For example, let $V = \mathbb{R}^2$ and $W = \mathbb{R}^3$. Let $\mathbf{w} = \{w_1, w_2, w_3\}$ and $\mathbf{v} = \{v_1, v_2\}$. Define T by $T(v_1) = 2w_1 + 3w_2 - w_3$ and $T(v_2) = w_1 + 5w_2 + w_3$. Then

$$[T]_{\mathbf{v}}^{\mathbf{w}} = \begin{bmatrix} 2 & 1 \\ 3 & 5 \\ -1 & 1 \end{bmatrix}$$

is the matrix of T with respect to the base pair $(\mathbf{v}, \mathbf{w})$.

Theorem 4.75 (**Composition of T and S as matrix multiplication**). *Let $S \in \mathbb{L}(U, V)$ and $T \in \mathbb{L}(V, W)$. Let $\mathbf{u} = (u_1, \ldots, u_p)$, $\mathbf{v} = (v_1, \ldots, v_q)$ and $\mathbf{w} = (w_1, \ldots, w_r)$ be bases for U, V, W. Then*

$$[TS]_{\mathbf{u}}^{\mathbf{w}} = [T]_{\mathbf{v}}^{\mathbf{w}}[S]_{\mathbf{u}}^{\mathbf{v}}.$$

Proof. For $j = 1, \ldots, p$, $(TS)(u_j) = T(S(u_j))$. Let $A = [T]_{\mathbf{v}}^{\mathbf{w}}$ and let $B = [S]_{\mathbf{u}}^{\mathbf{v}}$. Let $C = AB$. Thus,

$$(TS)(u_j) = T(S(u_j)) = T\left(\sum_{i=1}^{q} B(i,j)v_i\right).$$

By linearity of T we obtain

$$T\left(\sum_{i=1}^{q} B(i,j)v_i\right) = \sum_{i=1}^{q} B(i,j)T(v_i) = \sum_{i=1}^{q} B(i,j)\left(\sum_{t=1}^{r} A(t,i)w_t\right) =$$

$$\sum_{i=1}^{q}\sum_{t=1}^{r} B(i,j)A(t,i)w_t = \sum_{t=1}^{r}\left(\sum_{i=1}^{q} A(t,i)B(i,j)\right) w_t = \sum_{t=1}^{r} C(t,j)w_t.$$

Thus $(TS)(u_j) = \sum_{t=1}^{r} C(t,j)w_t$ for $j = 1, \ldots, p$ or $[TS]_{\mathbf{u}}^{\mathbf{w}} = [T]_{\mathbf{v}}^{\mathbf{w}}[S]_{\mathbf{u}}^{\mathbf{v}}$.

□

Theorem 4.75 has an interesting special case when $T \in \mathbb{L}(V, V)$.

Corollary 4.76 (**Change of basis for $T \in \mathbb{L}(V, V)$**). *Let $\mathbf{v} = (v_1, \ldots, v_q)$ and $\mathbf{w} = (w_1, \ldots, w_q)$ be ordered bases for the q-dimensional vector spaces V and W over the field $\mathbb{F}$ and let $T \in \mathbb{L}(V, V)$. Then*

$$[I]_{\mathbf{v}}^{\mathbf{w}}[T]_{\mathbf{v}}^{\mathbf{v}}[I]_{\mathbf{w}}^{\mathbf{v}} = [T]_{\mathbf{w}}^{\mathbf{w}}. \tag{4.77}$$

Proof. Apply Theorem 4.75.

□

REMARK 4.78 (**SIMILARITY OF MATRICES AND CHANGE OF BASES**). If T is the identity transformation, $T(x) = x$ for all $x \in V$, and $I_q \in \mathbf{M}_{n,n}(\mathbb{F})$ is the identity matrix, then 4.77 becomes $[I]_{\mathbf{v}}^{\mathbf{w}} I_q [I]_{\mathbf{w}}^{\mathbf{v}} = I_q$. Thus, if $S = [I]_{\mathbf{w}}^{\mathbf{v}}$ then S is nonsingular (invertible, unit) and $S^{-1} = [I]_{\mathbf{v}}^{\mathbf{w}}$. Given any basis $\mathbf{v} = (v_1, \ldots v_q)$ for V and any matrix $A \in \mathbf{M}_{n,n}(\mathbb{F})$, A and $\mathbf{v}$ define $T \in \mathbb{L}(V, V)$ by $A = [T]_{\mathbf{v}}^{\mathbf{v}}$. All $S \in \mathrm{GL}_n(\mathbb{F})$ can be interpreted as $[I]_{\mathbf{w}}^{\mathbf{v}}$ for selected bases (see discussion of this fact in remark 4.79). Thus, in matrix terms, 4.77 becomes $S^{-1}AS = B$ where $B = [T]_{\mathbf{w}}^{\mathbf{w}}$ with respect to the basis $\mathbf{w}$ which is defined by $S^{-1} = [I]_{\mathbf{v}}^{\mathbf{w}}$. When A and B are related by $S^{-1}AS = B$ for nonsingular S, they are called *similar* matrices.

REMARK 4.79 (**EQUIVALENCE OF BASES AND NONSINGULAR MATRICES**). Let $\mathcal{B} = \{\mathbf{u} \mid \mathbf{u} = (u_1, \ldots, u_q)$ a basis for V$\}$ be the set of all ordered bases for the q-dimensional vector space V over $\mathbb{F}$. Let $\mathrm{GL}_n(\mathbb{F})$ be all $n \times n$ nonsingular matrices over $\mathbb{F}$ (i.e., the general linear group). For a fixed basis $\mathbf{v} = (v_1, \ldots, v_q)$, the correspondence $\mathbf{u} \mapsto [I]_{\mathbf{u}}^{\mathbf{v}}$ (alternatively, we could work with $\mathbf{u} \mapsto [I]_{\mathbf{v}}^{\mathbf{u}}$) is a bijection between $\mathcal{B}$ and $\mathrm{GL}_n(\mathbb{F})$. In other words, given a fixed basis $\mathbf{v}$, we have that $P \in \mathrm{GL}_n(\mathbb{F})$ if and only if there is a basis $\mathbf{u}$ such that $P = [I]_{\mathbf{u}}^{\mathbf{v}}$.

Suppose $V = \mathbb{R}^3$ and $\mathbf{v} = (v_1, v_2, v_3) = ((1,0,0),(0,1,0),(0,0,1))$. Let

$$P = \begin{bmatrix} 1 & 0 & 1 \\ -1 & 1 & 0 \\ 0 & -1 & 2 \end{bmatrix} \equiv [I]_{\mathbf{u}}^{\mathbf{v}}$$

define a basis $\mathbf{u} = (u_1, u_2, u_3)$ where $u_1 = 1v_1 + (-1)v_2 + 0v_3 = (1,-1,0)$ and thus the transpose $(1,-1,0)^T = P^{(1)}$. Likewise, columns $P^{(2)}$ and $P^{(3)}$ define u_2 and u_3. It follows that

$$[I]_{\mathbf{v}}^{\mathbf{u}} = P^{-1} = \begin{bmatrix} 2/3 & -1/3 & -1/3 \\ 2/3 & 2/3 & -1/3 \\ 1/3 & 1/3 & 1/3 \end{bmatrix}$$

Because of this equivalence between matrices and linear transformations, most concepts in linear algebra have a "matrix version" and a "linear transformation" (or "operator") version. Going back and forth between these points of view can greatly simplify proofs.

DEFINITION 4.80 (**IMAGE AND KERNEL OF A LINEAR TRANSFORMATION**). Let $T \in \mathbb{L}(V, W)$ be a linear transformation. The set $\{T(x) \mid x \in V\}$ is a subspace of W called the *image* of T. It is denoted by image(T) or Im(T). The dimension, dim(Im(T)), is called the *rank*, $\rho(T)$, of T (see 3.119 and 3.120). The set $N(T) = \{x \mid x \in V, T(x) = 0_W\}$ where 0_W is the zero vector of W is called the *kernel* or *null space* of T. The $\dim(N(T)) = \eta(T)$ is called the *nullity* of T.

REMARK 4.81 (**REPRESENTING** $T(x) = y$ **IN MATRIX FORM:** $Ax = y$). Let $T \in \mathbb{L}(V, W)$, $\dim(V) = n$, $\dim(W) = m$. Let $\mathbf{v} = (v_1, \ldots, v_n)$ be a basis for V and $\mathbf{w} = (w_1, \ldots, w_m)$ a basis for W. If $x \in V$ write $x = \sum_{i=1}^{n} x_i v_i$. Let

$T(x) = T\left(\sum_{i=1}^{n} x_i v_i\right) = \sum_{j=1}^{n} x_j T(v_j)$. Defiine (a_{ij}) by $T(v_j) = \sum_{i=1}^{m} a_{ij} w_i$. Let $[x]^{\mathbf{v}}$ denote the column vector $[x_1, \ldots, x_n]^T$. Thus $A[x_1, \ldots, x_n]^T =$

$$(4.82) \qquad \begin{bmatrix} a_{11} & a_{12} & \ldots & a_{1n} \\ \vdots & \vdots & \vdots & \vdots \\ a_{m1} & a_{m2} & \ldots & a_{mn} \end{bmatrix} \begin{bmatrix} x_1 \\ \vdots \\ x_n \end{bmatrix} = [T]_{\mathbf{v}}^{\mathbf{w}} [x]^{\mathbf{v}} = [T(x)]^{\mathbf{w}}.$$

Note that multiplication of 4.82 on the left by a nonsingular $m \times m$ matrix P results in expressing $T(x)$ in a different basis:

$$PA[x_1, \ldots, x_n]^T = [I]_{\mathbf{w}}^{\mathbf{u}} [T(x)]^{\mathbf{w}} = [T(x)]^{\mathbf{u}}$$

since for any $\mathbf{w}$ and matrix P there is a unique basis $\mathbf{u} = (u_1, \ldots u_m)$ such that $P = [I]_{\mathbf{w}}^{\mathbf{u}} \in \mathrm{GL}_m(\mathbb{F})$ (see 4.79). As an example, suppose $[T]_{\mathbf{v}}^{\mathbf{w}} = A$ as follows:

$$(4.83) \qquad A = \begin{bmatrix} \overset{n_1}{\downarrow} & \overset{n_2}{\downarrow} & & \overset{n_3}{\downarrow} & & \\ 2 & 5 & 4 & -2 & 2 & 1 \\ 0 & 1 & 1 & 0 & -1 & 0 \\ 2 & 6 & 5 & 0 & -1 & 0 \end{bmatrix}.$$

By a sequence of elementary row operations

$$(4.84) \qquad P = R_{[1]+[3]} R_{(1/2)[3]} R_{[1]-(5/2)[2]} R_{(1/2)[1]} R_{[3]-[2]} R_{[3]-[1]}$$

we can construct a nonsingular P such that $PA = H$ is in Hermite canonical form:

$$PA = H = \begin{bmatrix} \overset{n_1}{\downarrow} & \overset{n_2}{\downarrow} & & \overset{n_3}{\downarrow} & & \\ 1 & 0 & \frac{-1}{2} & 0 & \frac{5}{2} & 0 \\ 0 & 1 & 1 & 0 & -1 & 0 \\ 0 & 0 & 0 & 1 & -1 & \frac{-1}{2} \end{bmatrix}$$

where $n_1 = 1$, $n_2 = 2$ and $n_3 = 4$ are the primary column indices (4.30) and

$$P = [I]_{\mathbf{w}}^{\mathbf{u}} = \begin{bmatrix} 0 & -3 & \frac{1}{2} \\ 0 & 1 & 0 \\ \frac{-1}{2} & \frac{-1}{2} & \frac{1}{2} \end{bmatrix}.$$

Thus, the matrix $H = [T]_{\mathbf{v}}^{\mathbf{u}}$ is the matrix of $T \in \mathbb{L}(V, W)$ with respect to the bases $\mathbf{v} = (v_1, \ldots, v_n)$ for V and $\mathbf{u} = (u_1, \ldots, u_m)$ for W. To obtain $\mathbf{u}$ explicitly in terms of $\mathbf{w}$, we need to compute

$$P^{-1} = [I]_{\mathbf{u}}^{\mathbf{w}} = \begin{bmatrix} 2 & 5 & -2 \\ 0 & 1 & 0 \\ 2 & 6 & 0 \end{bmatrix}$$

and thus obtain: $u_1 = 2w_1 + 2w_3$, $u_2 = 5w_1 + w_2 + 6w_3$ and $u_3 = -2w_1$.

REMARK 4.85 (**SOLVING EQUATIONS, RANK AND NULLITY**). We discuss the equation $T(x) = a$ where $T \in \mathbb{L}(V, W)$. Let $\mathbf{v} = (v_1, \ldots, v_n)$ be a basis for V and $\mathbf{w} = (w_1, \ldots, w_m)$ a basis for W. Using the notation of the previous discussion (4.81) we solve the equivalent matrix equation

$$A[x_1, \ldots, x_n]^T = [a_1, \ldots, a_m]^T$$

where $A \in \mathbf{M}_{m,n}(\mathbb{F})$ and $a = [a_1, \ldots, a_m]^T$ (see 4.82):

$$\begin{bmatrix} a_{11} & a_{12} & \ldots & a_{1n} \\ \vdots & \vdots & \vdots & \vdots \\ a_{m1} & a_{m2} & \ldots & a_{mn} \end{bmatrix} \begin{bmatrix} x_1 \\ \vdots \\ x_n \end{bmatrix} = \begin{bmatrix} a_1 \\ \vdots \\ a_m \end{bmatrix}. \tag{4.86}$$

The strategy is to multiply both sides of equation 4.86 by a matrix $P \in GL_m(\mathbb{F})$ to reduce A to its Hermite canonical form H. Thus, $PA[x_1, \ldots, x_n]^T = P[a_1, \ldots, a_m]^T$ becomes $H[x_1, \ldots, x_n]^T = [a'_1, \ldots, a'_m]^T$ where $PA = H$ and $P[a_1, \ldots, a_m]^T = [a'_1, \ldots, a'_m]^T$. Left unit equivalence preserves linear relations among columns (4.3) therefore

$$PA[x_1, \ldots, x_n]^T = P[a_1, \ldots, a_m]^T$$

if and only if

$$H[x_1, \ldots, x_n]^T = [a'_1, \ldots, a'_m]^T.$$

To help with the computation it is customary to form the *augmented* matrix of the system: $[A \,|\, a]$. For

$$A = \begin{bmatrix} 2 & 5 & 4 & -2 & 2 & 1 \\ 0 & 1 & 1 & 0 & -1 & 0 \\ 2 & 6 & 5 & 0 & -1 & 0 \end{bmatrix} \quad \text{and } a = \begin{bmatrix} 5 \\ -2 \\ -3 \end{bmatrix} \tag{4.87}$$

the augmented matrix is

$$[A|a] = \left[\begin{array}{cccccc|c} \overset{n_1}{\downarrow} & \overset{n_2}{\downarrow} & & \overset{n_3}{\downarrow} & & & \\ 2 & 5 & 4 & -2 & 2 & 1 & 5 \\ 0 & 1 & 1 & 0 & -1 & 0 & -2 \\ 2 & 6 & 5 & 0 & -1 & 0 & -3 \end{array}\right]. \tag{4.88}$$

By a sequence of elementary row operations (4.84) we construct a nonsingular P such that

$$P[A|a] = [PA \,|\, Pa] = [H|h] = \left[\begin{array}{cccccc|c} \overset{n_1}{\downarrow} & \overset{n_2}{\downarrow} & & \overset{n_3}{\downarrow} & & & \\ 1 & 0 & \frac{-1}{2} & 0 & \frac{5}{2} & 0 & \frac{9}{2} \\ 0 & 1 & 1 & 0 & -1 & 0 & -2 \\ 0 & 0 & 0 & 1 & -1 & \frac{-1}{2} & -3 \\ x_1 & x_2 & z_1 & x_3 & z_2 & z_3 & \end{array}\right]. \tag{4.89}$$

where H is the Hermite canonical form of A and $n_1 = 1$, $n_2 = 2$ and $n_3 = 4$ are the primary column indices. We can easily solve the equation

$$H[x_1, x_2, x_3, x_4, x_5, x_6]^T = [\frac{9}{2}, -2, -3]^T$$

by taking

$$[x_1, x_2, x_3, x_4, x_5, x_6] = [\frac{9}{2}, -2, 0, -3, 0, 0].$$

This same solution solves the original equation

$$A[x_1, x_2, x_3, x_4, x_5, x_6]^T = [5, -2, -3]^T.$$

where A and a are specified in 4.87. Thinking of A as a linear function from $\mathbb{F}^6$ to $\mathbb{F}^3$, the image of A is all of $\mathbb{F}^3$ and the rank of A is 3, the dimension of the image. Consider all vectors of the form (see 4.89 bottom line right)

$$z = [x_1, x_2, z_1, x_3, z_2, z_3].$$

Choose z_1, z_2, z_3 arbitrarily. Then choose $x_1 = (\frac{-1}{2})z_1 - (\frac{5}{2})z_2)$, $x_2 = -z_1 + z_2$ and $x_3 = z_2 + z_3/2$ to satisfy $Az = 0$ (for arbitrary $[z_1, z_2, z_3]$.) Thus, the null space, $\{z \mid Az = 0\}$ has dimension 3. The rank plus nullity, $\rho(A) + \eta(A) = 3 + 3 = 6 = \dim(\mathbb{F}^6)$. These ideas extend easily to the general case and show constructively why the rank plus nullity of a linear transformation on a vector space V equals $\dim(V)$.

Diagonal canonical forms – Smith form

For the rings $\mathbb{K}$ see 1.30. Start with a matrix $A \in \mathbf{M}_{m,n}(\mathbb{K})$ and reduce it to the column canonical form $AQ = H$ shown in 4.65 where $Q \in \mathrm{GL}_n(\mathbb{K})$. Next reduce the matrix H to row canonical form with $P \in \mathrm{GL}_m(\mathbb{K})$ to get $PAQ =$

(4.90)
$$\begin{bmatrix} d_1 & 0 & \cdots & 0 & 0 & \cdots & 0 \\ 0 & d_2 & \cdots & 0 & 0 & \cdots & 0 \\ \vdots & \vdots & \ddots & \vdots & \vdots & \vdots & \vdots \\ 0 & 0 & \cdots & d_{r-1} & 0 & \cdots & 0 \\ 0 & 0 & \cdots & 0 & d_r & \cdots & 0 \\ \mathbf{0} & \mathbf{0} & \cdots & \mathbf{0} & \mathbf{0} & \cdots & \mathbf{0} \end{bmatrix} = \left[\begin{array}{c|c} D_r & \Theta_{r,n-r} \\ \hline \Theta_{m-r,r} & \Theta_{m-r,n-r} \end{array}\right]$$

where $r = \rho(A)$. We will show that by further row and column operations the matrix PAQ of 4.90 can be reduced to a diagonal matrix in which $d_1|d_2|\cdots|d_r$ (i.e., the d_i form a divisibility chain).

DEFINITION 4.91 (**SMITH FORM, WEAK SMITH FORM**). A matrix $D \in \mathbf{M}_{m,n}(\mathbb{K})$ of the form shown in 4.90 is a *Smith form* if $r \geq 1$ and the diagonal elements $d_1, d_2, \ldots d_r$ form a divisibility chain: $d_1|d_2|\cdots|d_r$. If $d_1, d_2, \ldots d_r$ satisfy the weaker condition that $d_1|d_i$ for $i = 1, \ldots, r$ (i.e., d_1 divides all of the rest of the d_i) then we call D a *weak Smith form*.

Definition 4.92 (**Equivalence of matrices**). Matrices $A, B \in \mathbf{M}_{m,n}(\mathbb{K})$ are *left-right equivalent* if there exists $Q \in \mathrm{GL}_n(\mathbb{K})$ and $P \in \mathrm{GL}_m(\mathbb{K})$ such that $PAQ = B$. Note that left-right equivalence is an equivalence relation (1.35). Often, we refer to left-right equivalent matrices as just "equivalent matrices."

We will show that every $A \in \mathbf{M}_{m,n}(\mathbb{K})$ is equivalent (left-right) to a Smith form. If $\mathbb{K}$ is a field then 4.90 is a Smith form since all d_i are nonzero and hence units of the field. Thus, the case of interest will be when $\mathbb{K}$ is not a field. We first discuss the case $\rho(A) = 2$.

Remark 4.93 (**The case** $\rho(A) = 2$). Let $A \in \mathbf{M}_{m,n}(\mathbb{K})$ have rank $\rho(A) = 2$. The matrix PAQ of 4.90 becomes

$$D = \left[\begin{array}{c|c} D_2 & \Theta_{2,n-2} \\ \hline \Theta_{m-2,2} & \Theta_{m-2,n-2} \end{array}\right] = D_2 \oplus \Theta_{m-2,n-2}$$

where $D_2 = \begin{bmatrix} d_1 & 0 \\ 0 & d_2 \end{bmatrix}$. Since $\rho(D_2) = 2$, d_1 and d_2 are nonzero. Suppose $\mathbb{K}$ is a Euclidean domain but not a field (e.g., $\mathbb{Z}$ or $\mathbb{F}[x]$, see 1.30). Let $\gcd(d_1, d_2) = \delta$ and let $\mathrm{lcm}(d_1, d_2) = \lambda$ be the greatest common divisor and least common multiple of d_1 and d_2. From basic algebra, we have $\delta\lambda = d_1 d_2$ (1.31). Choose $s, t \in \mathbb{K}$ such that $sd_1 + td_2 = \delta$. Let $Q_2 = \begin{bmatrix} s & -d_2/\delta \\ t & d_1/\delta \end{bmatrix}$ and $P_2' = \begin{bmatrix} 1 & 1 \\ 0 & 1 \end{bmatrix}$. We have

$$P_2' D_2 Q_2 = \begin{bmatrix} 1 & 1 \\ 0 & 1 \end{bmatrix} \begin{bmatrix} d_1 & 0 \\ 0 & d_2 \end{bmatrix} \begin{bmatrix} s & -d_2/\delta \\ t & d_1/\delta \end{bmatrix}$$

where

$$\begin{bmatrix} 1 & 1 \\ 0 & 1 \end{bmatrix} \begin{bmatrix} d_1 & 0 \\ 0 & d_2 \end{bmatrix} = \begin{bmatrix} d_1 & d_2 \\ 0 & d_2 \end{bmatrix}$$

and

$$\begin{bmatrix} d_1 & d_2 \\ 0 & d_2 \end{bmatrix} \begin{bmatrix} s & -d_2/\delta \\ t & d_1/\delta \end{bmatrix} = \begin{bmatrix} \delta & 0 \\ td_2 & \lambda \end{bmatrix} \tag{4.94}$$

Finally, noting that $\delta \mid td_2$ (in fact, $\delta \mid d_2$), applying the elementary row matrix $R_{[2]-c[1]}$ to 4.94 where $c = td_2/\delta$ results in

$$P_2 D_2 Q_2 = \begin{bmatrix} \delta & 0 \\ 0 & \lambda \end{bmatrix} = \begin{bmatrix} d_1 \wedge d_2 & 0 \\ 0 & d_1 \vee d_2 \end{bmatrix} = \hat{D}_2 \tag{4.95}$$

where $P_2 = R_{[2]-c[1]} P_2'$ and $Q_2 = \begin{bmatrix} s & -d_2/\delta \\ t & d_1/\delta \end{bmatrix}$. We use the notation $d_1 \wedge d_2 = \gcd(d_1, d_2)$ and $d_1 \vee d_2 = \mathrm{lcm}(d_1, d_2)$ and the fact that $d_1 d_2 / \gcd(d_1, d_2) = \mathrm{lcm}(d_1, d_2)$ (1.31). Note that the diagonal matrix $\hat{D}_2$ has rank 2, and $\hat{D}_2(1,1)$ divides $\hat{D}_2(2,2)$. $\hat{D}_2$ is a Smith form for D_2.

Lemma 4.96 (**Weak Smith form**). *Let $A \in \mathbf{M}_{m,n}(\mathbb{K})$ with $\rho(A) = r > 0$. There exists $Q \in \mathrm{GL}_n(\mathbb{K})$ and $P \in \mathrm{GL}_m(\mathbb{K})$ such $PAQ = \left[\begin{array}{c|c} D_r & \Theta_{r,n-r} \\ \hline \Theta_{m-r,r} & \Theta_{m-r,n-r} \end{array}\right]$ where $D_r = \mathrm{diag}(\mathrm{d}_1, \ldots, \mathrm{d}_r)$ and $d_1 | d_i$ for $i = 1 \ldots r$ (i.e., weak Smith form).*

Proof. The proof is by induction on r. The case $r = 1$ is trivial. The case $r = 2$ was shown in remark 4.93. Let $r > 2$ and assume the lemma is true for $r - 1$. We can choose P and Q such that

$$PAQ = \left[\begin{array}{c|c} D_r & \Theta_{r,n-r} \\ \hline \Theta_{m-r,r} & \Theta_{m-r,n-r} \end{array}\right] \tag{4.97}$$

as in 4.90 where $D_r = \mathrm{diag}(\mathrm{d}_1, \ldots \mathrm{d}_r)$. By the induction hypothesis, we can further apply left-right multiplications by nonsingular matrices (or, equivalently, row and column operations) such that, using the same notation of 4.97, $d_1 | d_i$ for $i = 1, \ldots, r - 1$. Next, by applying the result for $r = 2$ (4.93), we can construct $\hat{P} = P_2 \oplus_{\{1,r\}}^{\{1,r\}} I_{m-2}$ and $\hat{Q} = Q_2 \oplus_{\{1,r\}}^{\{1,r\}} I_{n-2}$ (notation 3.25) such that

$$\hat{P} D_r \hat{Q} = \hat{D}_r = \mathrm{diag}(\mathrm{d}_1 \wedge \mathrm{d}_r, \mathrm{d}_2, \ldots \mathrm{d}_{r-1}, \mathrm{d}_1 \vee \mathrm{d}_r)$$

where now, $\hat{D}_r(1,1) | \hat{D}_r(i,i)$ for $i = 1 \ldots r$. This completes the proof.

□

Theorem 4.98 (**Smith form**). *Let $A \in \mathbf{M}_{m,n}(\mathbb{K})$, $\rho(A) = r > 0$. There exists $Q \in \mathrm{GL}_n(\mathbb{K})$ and $P \in \mathrm{GL}_m(\mathbb{K})$ such $PAQ = \left[\begin{array}{c|c} D_r & \Theta_{r,n-r} \\ \hline \Theta_{m-r,r} & \Theta_{m-r,n-r} \end{array}\right]$ where $D_r = \mathrm{diag}(\mathrm{d}_1, \ldots, \mathrm{d}_r)$ and $d_1 | d_2 | \cdots | d_r$. Thus, every $A \in \mathbf{M}_{m,n}(\mathbb{K})$ is equivalent to a Smith form. Alternatively, any $A \in \mathbf{M}_{m,n}(\mathbb{K})$ can be transformed into a Smith form by elementary row and column operations (4.33).*

Proof. By lemma 4.96, there exists $\tilde{P}$ and $\tilde{Q}$ such that

$$\tilde{P} A \tilde{Q} = \left[\begin{array}{c|c} \tilde{D}_r & \Theta_{r,n-r} \\ \hline \Theta_{m-r,r} & \Theta_{m-r,n-r} \end{array}\right] \quad \text{(weak Smith form)}$$

where $\tilde{D}_r = \mathrm{diag}(\tilde{\mathrm{d}}_1, \tilde{\mathrm{d}}_2, \ldots, \tilde{\mathrm{d}}_r)$ and $\tilde{d}_1 | \tilde{d}_i$ for $i = 1, \ldots, r$.

Thus, we need only show that there are $\hat{P} \in \mathrm{GL}_r(\mathbb{K})$ and $\hat{Q} \in \mathrm{GL}_r(\mathbb{K})$ such that $\hat{P} \tilde{D}_r \hat{Q}$ is in Smith form. The proof is by induction. The case $r = 1$ is trivial. The case $r = 2$ was shown in remark 4.93. Note, in particular, equation 4.95. Let $r > 2$ and assume the theorem is true for $r - 1$. The induction hypothesis applied to $D'_{r-1} = \tilde{D}_r(1|1) = \mathrm{diag}(\tilde{\mathrm{d}}_2, \ldots, \tilde{\mathrm{d}}_r)$ implies there exists $P' \in \mathrm{GL}_{r-1}(\mathbb{K})$ and $Q' \in \mathrm{GL}_{r-1}(\mathbb{K})$ such that $P' D'_{r-1} Q' = \mathrm{diag}(\mathrm{d}_2, \mathrm{d}_3, \ldots, \mathrm{d}_r)$ where $d_2 | d_3 | \cdots | d_r$. Note that since $\tilde{d}_1$ divides all entries of $D'_{r-1} = \mathrm{diag}(\tilde{\mathrm{d}}_2, \ldots, \tilde{\mathrm{d}}_r)$, $\tilde{d}_1$ also divides all entries of $P' D'_{r-1} Q'$ (easily seen for multiplication by elementary row and column operations) and hence divides d_2. Thus, setting $\tilde{d}_1 \equiv d_1$ we have $d_1 | d_2 | \cdots | d_r$. Taking $\hat{P} = (1) \oplus P'$ and $\hat{Q} = (1) \oplus Q'$ gives

$\hat{P}\tilde{D}_r\hat{Q} = \mathrm{diag}(\mathrm{d}_1, \mathrm{d}_2, \ldots, \mathrm{d}_\mathrm{r})$ which is a Smith form.

□

COROLLARY 4.99 (**PAIRWISE RELATIVELY PRIME DIAGONAL ENTRIES**). *Let*

$$\tilde{D} = \mathrm{diag}(\tilde{\mathrm{d}}_1, \tilde{\mathrm{d}}_2, \ldots, \tilde{\mathrm{d}}_\mathrm{n}) \in \mathbf{M}_{\mathrm{n,n}}(\mathbb{K}), \quad \mathrm{n} > 1,$$

and suppose that $\gcd(\tilde{d}_i, \tilde{d}_j) = 1$ *for* $1 \le i < j \le n$. *Then*

$$D = (1, \ldots, 1, (\tilde{d}_1\tilde{d}_2 \cdots \tilde{d}_n))$$

is a Smith form for $\tilde{\mathrm{D}}$.

PROOF. We use the notation $a \wedge b = \gcd(a, b)$ and $a \vee b = \mathrm{lcm(a, b)}$ (1.31). The case $n = 2$ follows from 4.95: $\tilde{D} = \mathrm{diag}(\tilde{\mathrm{d}}_1, \tilde{\mathrm{d}}_2)$ is equivalent to

$$D = \begin{bmatrix} \tilde{d}_1 \wedge \tilde{d}_2 & 0 \\ 0 & \tilde{d}_1 \vee \tilde{d}_2 \end{bmatrix} = \begin{bmatrix} 1 & 0 \\ 0 & \tilde{d}_1\tilde{d}_2 \end{bmatrix}.$$

Assume the case $n - 1$. Then $\tilde{D} = \mathrm{diag}(\tilde{\mathrm{d}}_1, \tilde{\mathrm{d}}_2, \ldots, \tilde{\mathrm{d}}_\mathrm{n})$ is equivalent to $D' = \mathrm{diag}(1, \ldots, 1, (\tilde{\mathrm{d}}_1\tilde{\mathrm{d}}_2 \cdots \tilde{\mathrm{d}}_{\mathrm{n}-1}), \tilde{\mathrm{d}}_\mathrm{n})$. Using the case 2 again on the last two entries we get the result. □

CHAPTER 5

Similarity and equivalence

In this chapter we focus on the Euclidean domains (1.21) $\mathbb{K} \in \{\mathbb{Z}, \mathbb{F}[x]\}$, $\mathbb{F}$ a field as specified in remark 1.30.

Determinantal divisors and related invariants

DEFINITION 5.1 (**k^{TH} ORDER DETERMINANTAL DIVISOR**). Let $A \in \mathbf{M}_{m,n}(\mathbb{K})$ where $\mathbb{K}$ is a Euclidean domain ($\mathbb{K} \in \{\mathbb{Z}, \mathbb{F}[x]\}$, $\mathbb{F}$ a field as specified in remark 1.30). Let $1 \leq k \leq \min(m, n)$. Let f_k denote a greatest common divisor of all $k \times k$ subdeterminants of A:

$$f_k = \gcd\{\det(A_\omega^\gamma) \mid \omega \in \mathrm{SNC}(k, m), \gamma \in \mathrm{SNC}(k, n)\} \tag{5.2}$$

where $\mathrm{SNC}(k, m)$ denotes the strictly increasing functions from $\underline{k} \to \underline{m}$ and A_ω^γ is the submatrix of A with rows selected by ω and columns by γ (2.52). We call f_k a kth *order determinantal divisor* of A; it is determined up to units in $\mathbb{K}$. We define $f_0 = 1$ (the multiplicative identity in $\mathbb{K}$).

DEFINITION 5.3 (**DETERMINANTAL DIVISOR SEQUENCES**). Let $A \in \mathbf{M}_{m,n}(\mathbb{K})$ with $\mathbb{K}$ as in remark 1.30. Let f_k, $0 \leq k \leq \min(m, n)$, be as in definition 5.1. Define

$$(f_0^A, f_1^A, \ldots, f_{\min(m,n)}^A)$$

to be a *sequence of determinantal divisors* of A. Let $\rho(A)$ (3.119) be the rank of A. Define

$$(f_0^A, f_1^A, \ldots, f_{\rho(A)}^A)$$

to be a *maximal sequence of nonzero determinantal divisors* of A.

REMARK 5.4 (**EXAMPLE OF DETERMINANTAL DIVISOR SEQUENCES**). To save space, we sometimes write $\det X = |X|$. Let $A \in \mathbf{M}_{3,4}(\mathbb{Z})$.

$$A = \begin{bmatrix} 0 & 4 & 6 & 2 \\ 8 & 2 & 10 & 8 \\ 2 & 0 & 4 & 4 \end{bmatrix} \tag{5.5}$$

By definition, $f_0^A = 1$. Obviously, $f_1^A = 2$ or -2. Let's choose $f_1^A = 2$, using the canonical SDR for associates for $\mathbb{Z}$ (4.36). Since all 2×2 subdeteriminants of A have even entries, all such subdeterminants are divisible by 4. The determinant of $A[2, 3 \mid 1, 2]$ is -4. Thus, $f_2^A = 4$, again choosing from the canonical SDR.

Clearly, $\rho(A) = 3$ since $\det A[1,2,3 \mid 1,2,3] = -72 \neq 0$. There are four possible 3×3 subdeterminants: $|A^{\gamma}|$ where

$$\gamma = (1,2,3),\ (1,2,4),\ (1,3,4) \text{ or } (2,3,4).$$

Note that columns $A^{(2)} + A^{(4)} = A^{(3)}$. This implies that $|A^{\gamma}| = 0$ if $\gamma = (2,3,4)$. Check that $|A^{(1,2,3)}| = -72$, $|A^{(1,2,4)}| = -72$, $|A^{(1,3,4)}| = -72$ and, thus, $f_3^A = 72$. We have,

$$(f_0^A, f_1^A, f_2^A, f_3^A) = (1, 2, 4, 72)$$

is both a sequence of determinantal divisors and a maximal sequence of nonzero determinantal divisors of A.

Recall the definition of left-right equivalence (or just equivalence) of matrices 4.92. By corollary 3.121 we know that if A and B are equivalent, $A = PBQ$, then their ranks are equal: $\rho(A) = \rho(B)$.

Lemma 5.6 (**Determinantal divisors of equivalent matrices**). *Let $A, B \in \mathbf{M}_{m,n}(\mathbb{K})$ and let $Q \in \mathrm{GL}_n(\mathbb{K})$, $P \in \mathrm{GL}_m(\mathbb{K})$ be such that $A = PBQ$. Let $r = \rho(A) = \rho(B)$. Then the determinantal divisor sequences satisfy*

$$(f_0^A, f_1^A, \ldots, f_r^A) = (f_0^B, u_1 f_1^B, \ldots, u_r f_r^B) \tag{5.7}$$

where u_i, $1 \le i \le r$, are units in $\mathbb{K}$. If the determinantal divisors f_i^A and f_i^B, $1 \le i \le r$, come from the same SDR for associates in $\mathbb{K}$ (4.36) then

$$(f_0^A, f_1^A, \ldots, f_r^A) = (f_0^B, f_1^B, \ldots, f_r^B). \tag{5.8}$$

Proof. If $A = \Theta_{m,n}$ then 5.7 and 5.8 are trivial: $(1) = (1)$. Assume $A \neq \Theta_{m,n}$. We use Cauchy-Binet, corollary 3.107 (equation 3.109). Let $1 \le k \le r$ and choose $g \in \mathrm{SNC(k, m)}$ and $h \in \mathrm{SNC(k, n)}$. From 3.109 with $X = PB \in \mathbf{M}_{m,n}(\mathbb{K})$ we have

$$\det(X_g^h) = \det((PB)_g^h) = \det(P_g B^h) = \sum_{f \in \mathrm{SNC}(k,m)} \det(P_g^f)\det(B_f^h) \tag{5.9}$$

where $X_g^h \in \mathbf{M}_{k,k}(\mathbb{K})$, $P_g \in \mathbf{M}_{k,m}(\mathbb{K})$, $B^h \in \mathbf{M}_{m,k}(\mathbb{K})$ and P_g^f, $B_f^h \in \mathbf{M}_{k,k}(\mathbb{K})$. Note that the kth $(1 \le k \le r)$ determinantal divisor $f_k^B \neq 0$ divides $\det(B_f^h)$ for all $f \in \mathrm{SNC}(k, m)$. Hence f_k^B divides $\det(X_g^h)$ for all $g \in \mathrm{SNC(k, m)}$ and $h \in \mathrm{SNC(k, n)}$. Thus, f_k^B divides f_k^X. But $P^{-1}X = B$ so the same argument yields f_k^X divides f_k^B and hence for $1 \le k \le r$, $f_k^X = u_k f_k^B$ where $u_k \in \mathbb{K}$ is a unit. A similar argument shows that if $Y = BQ$ then $f_k^Y = v_k f_k^B$ for $1 \le k \le r$, where $v_k \in \mathbb{K}$ is a unit. Applying these two results to $A = PBQ$ completes the proof. □

Lemma 5.10 (**Divisibility and determinantal divisors**). *Let $A \in \mathbf{M}_{m,n}(\mathbb{K})$ and let $(f_0^A, f_1^A, \ldots, f_r^A)$ be a maximal sequence of nonzero determinantal divisors of A where $r = \rho(A)$. Then $f_k | f_{k+1}$ for $0 \le k < \rho(A)$.*

PROOF. Let $A \in \mathbf{M}_{m,n}(\mathbb{K})$, $g \in \mathrm{SNC}(k+1, m)$, $h \in \mathrm{SNC}(k+1, n)$ for $1 \le k < r$. From corollary 3.72, simple Laplace expansion by the ith row of the $(k+1) \times (k+1)$ matrix A_g^h gives

$$\det(A_g^h) = (-1)^i \sum_{j=1}^{k+1} (-1)^j A_g^h(i,j) \det(A_g^h(i \mid j)). \tag{5.11}$$

Since $A_g^h(i \mid j)$ is a $k \times k$ matrix, $f_k^A \mid \det(A_g^h(i \mid j))$ for $1 \le j \le k+1$. Thus, $f_k^A \mid \det(A_g^h)$ for all choices of g and h and hence $f_k^A \mid f_{k+1}^A$. This completes the proof. □

DEFINITION 5.12 (**SEQUENCE OF INVARIANT FACTORS**). Let $A \in \mathbf{M}_{m,n}(\mathbb{K})$ and let $(f_0^A, f_1^A, \ldots, f_r^A)$ be a maximal sequence of nonzero determinantal divisors of A where $r = \rho(A)$. The sequence $(q_1^A, \ldots, q_r^A)$ where $q_k^A = f_k^A / f_{k-1}^A$, $k = 1, \ldots, r$, is called a *sequence of invariant factors* of the sequence $(f_0^A, f_1^A, \ldots, f_r^A)$. From lemma 5.6, the sequences of invariant factors of equivalent matrices are the same up to units in $\mathbb{K}$.

THEOREM 5.13 (**SMITH FORM AND INVARIANT FACTORS**). *Let $A \in \mathbf{M}_{m,n}(\mathbb{K})$, $\rho(A) = r > 0$. Let* $D = PAQ = \left[\begin{array}{c|c} D_r & \Theta_{r,n-r} \\ \hline \Theta_{m-r,r} & \Theta_{m-r,n-r} \end{array}\right]$ *where $D_r = \mathrm{diag}(d_1, \ldots, d_r)$ and $d_1 | d_2 | \cdots | d_r$ be a Smith form of A (4.98). Then $(q_1^A, \ldots, q_r^A)$ where $q_i^A = d_i$, $i = 1, \ldots, r$, is a sequence of invariant factors for A. Thus, the sequence of invariant factors satisfies $q_1^A | q_2^A | \cdots | q_{r-1}^A | q_r^A$.*

PROOF. Recall lemma 5.6 which states that equivalent matrices have the same determinantal divisor sequences (up to units). The special structure of D implies that for $k = 1, \ldots, r$

$$f_k^A = \gcd\{\prod_{i=1}^{k} d_{\gamma(i)} \mid \gamma \in \mathrm{SNC}(k, r)\} \tag{5.14}$$

where $\mathrm{SNC}(k, r)$ denotes the strictly increasing functions from $\underline{k} \mapsto \underline{r}$. The fact that $d_1 | d_2 | \cdots | d_r$ forms a divisibility chain implies that $\prod_{i=1}^k d_i \mid \prod_{i=1}^k d_{\gamma(i)}$ for all $\gamma \in \mathrm{SNC}(k, r)$. Thus, $f_k^A = \prod_{i=1}^k d_i$, $i = 1, \ldots, r$, and $q_k^A = f_k^A / f_{k-1}^A = d_k$ define a sequence of determinantal divisors and a corresponding sequence of invariant factors of A. Note $d_1 | d_2 | \cdots | d_{r-1} | d_r$ implies $q_1^A | q_2^A | \cdots | q_{r-1}^A | q_r^A$. This completes the proof.

□

COROLLARY 5.15 (**INVARIANTS WITH RESPECT TO EQUIVALENCE**). *Let $A, B \in \mathbf{M}_{m,n}(\mathbb{K})$, $\rho(A) = r > 0$. If A and B are equivalent matrices (4.92) then the following sequences are the same up to units:*

(1) **Determinantal divisors** $(f_0^A, f_1^A, \ldots, f_r^A)$ *and* $(f_0^B, f_1^B, \ldots, f_r^B)$

(2) **Invariant factors** $(q_1^A, \ldots, q_r^A)$ *and* $(q_1^B, \ldots, q_r^B)$

(3) **Smith form diagonal entries** $(d_1^A, \ldots, d_r^A)$ *and* $(d_1^B, \ldots, d_r^B)$.

If any one of the sequence pairs (1), (2) *or* (3) *are the same up to units then* A *and* B *are equivalent.*

Proof. By theorem 5.13, the Smith form diagonal entries are the same as the invariant factors up to units. Likewise, by definition 5.12, the sequence of determinantal divisors determines the sequence of invariant factors and conversly. By lemma 5.6, if A and B are equivalent matrices then they have the same sequences of determinantal divisors up to units. In particular, (3) implies that A and B are equivalent to the same Smith form (up to units) and are thus equivalent to each other.

□

Definition 5.16 (**Elementary divisors**). Let $A \in \mathbf{M}_{m,n}(\mathbb{K})$, $\rho(A) = r > 0$. Let $(q_1^A, \ldots, q_r^A)$, $i = 1, \ldots, r$, be a sequence of invariant factors for A. By a set of *distinct* primes we mean a set of primes $X = \{p_1, p_2, \ldots, p_s\}$, $|X| = s$. Assume that X is chosen to be minimal in the sense that $p \in X$ if and only if there exists an invariant factor q_i^A, $i = 1, \ldots, r$, such that $p|q_i^A$. Factor all of the q_i^A into prime factors as follows:

$$
\begin{array}{rcl} q_1^A &=& p_1^{e_{11}} p_2^{e_{12}} \quad \cdots \quad p_s^{e_{1s}} \\ q_2^A &=& p_1^{e_{21}} p_2^{e_{22}} \quad \cdots \quad p_s^{e_{2s}} \\ \vdots & & \qquad\quad \vdots \\ q_r^A &=& p_1^{e_{r1}} p_2^{e_{r2}} \quad \cdots \quad p_s^{e_{rs}}. \end{array} \tag{5.17}
$$

A multiset (1.32) of *elementary divisors* of A is

$$\{p_i^{e_{ij}} \mid (i,j) \in \underline{r} \times \underline{s},\ e_{ij} > 0\}.$$

The elementary divisors are determined up to multiplication by units.

Remark 5.18 (**Multisets (1.32) of elementary divisors**). Let $A \in \mathbf{M}_{m,n}(\mathbb{K})$, $\rho(A) = r > 0$. Both sequences $(f_0^A, \ldots, f_r^A)$ and $(q_1^A, \ldots, q_r^A)$ form divisibility chains. Thus, these sequences can be reconstructed from the corresponding multisets $\{f_0^A, \ldots, f_r^A\}$ and $\{q_1^A, \ldots, q_r^A\}$ by sorting the multisets in order by divisibility. The situation for elementary divisors is similar but requires more discussion. Referring to equation 5.17, the divisibility condition, $q_1^A | q_2^A | \cdots | q_{r-1}^A | q_r^A$, implies that each sequence of exponents is weakly increasing: $e_{1j} \le e_{2j} \le \cdots \le e_{rj}$, $1 \le j \le s$. Thus, knowing the multiset $\{e_{1j}, e_{2j}, \cdots, e_{rj}\}$ for a particular j is enough to reconstruct the sequence of exponents for that j. Note also that if you know the rank r, then knowing the multiset of nonzero e_{ij} (corresponding to the elementary divisors) is enough to determine the entire sequence $e_{1j} \le e_{2j} \le \cdots \le e_{rj}$, $1 \le j \le s$. Likewise, the sequences of powers of individual primes are determined by their multisets:

$$
(5.19) \qquad \begin{matrix} p_1 & p_2 & & p_s \\ \begin{bmatrix} p_1^{e_{11}} \\ p_1^{e_{21}} \\ \vdots \\ p_1^{e_{r1}} \end{bmatrix} & \begin{bmatrix} p_2^{e_{12}} \\ p_2^{e_{22}} \\ \vdots \\ p_2^{e_{r2}} \end{bmatrix} & \begin{matrix} \cdots \\ \cdots \\ \vdots \\ \cdots \end{matrix} & \begin{bmatrix} p_s^{e_{1s}} \\ p_s^{e_{2s}} \\ \vdots \\ p_s^{e_{rs}} \end{bmatrix} \end{matrix}.
$$

Thus, the multiset $\{p_2^{e_{12}}, p_2^{e_{22}}, \ldots, p_2^{e_{r2}}\}$ determines the second column in 5.19. In fact, if you know r, the multiset of all $p_2^{e_{i2}}$ where $e_{i2} > 0$ determines the second column. Thus, the second column is determined by the elementary divisors of the form $p_2^{e_{i2}}$. Knowing the rank r, the multiset of all elementary divisors can be broken down (by determining the primes p_i) into the individual multisets corresponding to the primes p_i, $i = 1, \ldots, s$. See remark 5.21 for an example.

LEMMA 5.20 (**ELEMENTARY DIVISORS DETERMINE INVARIANT FACTORS**). *Let $A \in \mathbf{M}_{m,n}(\mathbb{K})$, $\rho(A) = r > 0$. Let $(q_1^A, \ldots, q_r^A)$, $i = 1, \ldots, r$, be a sequence of invariant factors for A. Let the elementary divisors be specified as in 5.17. Then r together with the multiset (1.32) of elementary divisors*

$$\{p_i^{e_{ij}} \mid 1 \le i \le r, 1 \le j \le s, e_{ij} > 0\}$$

determines the invariant factors and determinantal divisors (up to units). Thus, two matrices in $\mathbf{M}_{m,n}(\mathbb{K})$are equivalent if and only if they have the same multiset of elementary divisors and same rank (see 5.15).

PROOF. The idea for the proof is developed in remark 5.18. An example is given in remark 5.21. □

REMARK 5.21 (**EXAMPLE OF ELEMENTARY DIVISORS TO INVARIANT FACTORS**). Suppose $r = 6$ and the multiset of elementary divisors is

$$X = \{2, 2, 3, 3, 4, 4, 5, 5, 7, 7, 9, 9, 9, 25, 49\}.$$

The multisets X_2, X_3, X_5, X_7 associated with the primes 2, 3, 5, 7 are

$$(5.22) \qquad \{2^1, 2^1, 2^2, 2^2\},\ \{3^1, 3^1, 3^2, 3^2, 3^2\},\ \{5^1, 5^1, 5^2\},\ \{7^1, 7^1, 7^2\}.$$

Sorting each of these sets into numerical order (as they are already listed) and prefixing the number of ones (e.g., in the form 2^0, 3^0, 5^0, 7^0) needed to make $r = 6$ items in the sorted list gives the columns of the following array which represents the factorization of the invariant factors into powers of primes (as in 5.17):

$$q_1^A = 2^0 \quad 3^0 \quad 5^0 \quad 7^0$$
$$q_2^A = 2^0 \quad 3^1 \quad 5^0 \quad 7^0$$
$$q_3^A = 2^1 \quad 3^1 \quad 5^0 \quad 7^0$$
$$(5.23) \qquad q_4^A = 2^1 \quad 3^2 \quad 5^1 \quad 7^1 \; .$$
$$q_5^A = 2^2 \quad 3^2 \quad 5^1 \quad 7^1$$
$$q_6^A = 2^2 \quad 3^2 \quad 5^2 \quad 7^2$$

A recursive approach is to construct the multisets X_2, X_3, X_5, X_7 as in 5.22 and remove the highest powers of each prime to get $q_6^A = 2^2 3^2 5^2 7^2$ and new sets X_2', X_3', X_5', X_7':

$$(5.24) \qquad \{2^1, 2^1, 2^2\},\ \{3^1, 3^1, 3^2, 3^2\},\ \{5^1, 5^1\},\ \{7^1, 7^1\}.$$

Proceed recursively from 5.24 to get $q_5^A, \ldots, q_2^A$. The remaining invariant factor, $q_1^A = 1$ is determined by knowing the rank $r = 6$.

Equivalence vs. similarity

Matrices $X, Y \in \mathbf{M}_{m,n}(\mathbb{K})$ are *equivalent* if there exists $Q \in \mathrm{GL}_n(\mathbb{K})$ and $P \in \mathrm{GL}_m(\mathbb{K})$ such that $PXQ = Y$ (4.92). In this section we take $m = n$ and $\mathbb{K} = \mathbb{F}[x]$ where $\mathbb{F}$ is a field and $\mathbb{F}[x]$ is the ring of polynomials with coefficients in $\mathbb{F}$. See remark 1.30. As previously, for $\mathbf{M}_{n,n}(\mathbb{K})$, we write $\mathbf{M}_n(\mathbb{K})$. Note that $\mathbf{M}_n(\mathbb{F})$ is a subring of $\mathbf{M}_n(\mathbb{F}[x])$. As an example of an element of $\mathbf{M}_n(\mathbb{F}[x])$, let

$$(5.25) \qquad P = \begin{bmatrix} x^2/3 & x^3 - x^2/2 \\ 2x^3 + 2/5 & 2x - 3 \end{bmatrix}$$

be in $\mathbf{M}_2(\mathbb{F}[x])$ where $\mathbb{F} = \mathbb{Q}$, the rational numbers. Note that

$$(5.26) \qquad P = P_3 x^3 I + P_2 x^2 I + P_1 x^1 I + P_0 x^0 I$$

where I is the 2×2 identity matrix, $x^t I = \mathrm{diag}(\mathrm{x^t}, \mathrm{x^t})$, and

$$P_3 = \begin{bmatrix} 0 & 1 \\ 2 & 0 \end{bmatrix} \quad P_2 = \begin{bmatrix} 1/3 & -1/2 \\ 0 & 0 \end{bmatrix} \quad P_1 = \begin{bmatrix} 0 & 0 \\ 0 & 2 \end{bmatrix} \quad P_0 = \begin{bmatrix} 0 & 0 \\ 2/5 & -3 \end{bmatrix}.$$

Recall delta notation, $\delta(\text{Statement}) = 1$ if "Statement" is true, 0 otherwise (1.1). The matrix $P_k \in \mathbf{M}_2(\mathbb{F})$ is defined by $P_k(i,j) = \mathrm{coeff}(x^k, P(i,j))$, $(i,j) \in \underline{2} \times \underline{2}$, where $\mathrm{coeff}(x^k, P(i,j))$ denotes the coefficient of x^k in the polynomial $P(i,j)$.

Definition 5.27 (**Canonical presentation**). Let $P \in \mathbf{M}_n(\mathbb{F}[x])$ and let $x^k I = \mathrm{diag}(x^k, \ldots, x^k)$ where I is the $n \times n$ identity. For each $0 \le k \le m$, define $P_k \in \mathbf{M}_n(\mathbb{F})$ by $P_k(i,j) = \mathrm{coeff}(x^k, P(i,j)), (i,j) \in \underline{n} \times \underline{n}$, where $\mathrm{coeff}(x^k, P(i,j))$ denotes the coefficient of x^k in the polynomial $P(i,j)$. Let $m = \max\{\mathrm{degree}(P(i,j)) \,|\, (i,j) \in \underline{n} \times \underline{n}\}$ (see 5.25 and 5.26). The *canonical* *presentation* of P is

$$P = \sum_{k=0}^{m} P_k x^k I. \tag{5.28}$$

Remark 5.29 (**Uniqueness of canonical presentation**). Note that the canonical presentation of a matrix $P \in \mathbf{M}_n(\mathbb{F}[x])$ is unique in the sense that if $A_k \in \mathbf{M}_n(\mathbb{F})$, $0 \le k \le m$, then (see 5.28)

$$P = \sum_{k=0}^{m} P_k x^k I = \sum_{k=0}^{m} A_k x^k I \;\Rightarrow\; A_k = P_k,\; 0 \le k \le m. \tag{5.30}$$

Identity 5.30 is evident when $m = 0$ and can be proved by induction on m. Observe that $x^k I$ commutes with every matrix in $\mathbf{M}_n(\mathbb{F}[x])$. Using the concept of a module (1.24), the additive group of the ring $\mathbf{M}_n(\mathbb{F}[x])$ is a left R-module where $R = \mathbf{M}_n(\mathbb{F})$. Verifying the axioms of definition 1.24 is trivial since $R = \mathbf{M}_n(\mathbb{F})$ is a subring of $\mathbf{M}_n(\mathbb{F}[x])$. The set $X = \{x^k I : k = 0, 1, \ldots\}$ is a *infinite module basis* for this module.

We next define functions ρ_A and λ_A (called *right and left evaluation*) from $\mathbf{M}_n(\mathbb{F}[x])$ to itself. Right evaluation ρ_A turns out to be *linear* for the left R-module in the sense that $\rho_A(\alpha P + \beta Q) = \alpha\rho_A(P) + \beta\rho_A(Q)$ for $\alpha, \beta \in \mathbf{M}_n(\mathbb{F})$ and $P, Q \in \mathbf{M}_n(\mathbb{F}[x])$. A symmetric result holds for left evaluation λ_A which is linear for the right R-module (i.e. $\lambda_A(P\alpha + Q\beta) = \lambda_A(P)\alpha + \lambda_A(Q)\beta$).

Definition 5.31 (**Evaluation by $A \in \mathbf{M}_n(\mathbb{F})$**). Let $A \in \mathbf{M}_n(\mathbb{F})$ and let $P \in \mathbf{M}_n(\mathbb{F}[x])$ have canonical presentation $P = \sum_{k=0}^{m} P_k x^k I$. Define functions, *right and left evaluation*, ρ_A, λ_A from $\mathbf{M}_n(\mathbb{F}[x])$ to $\mathbf{M}_n(\mathbb{F})$ by

$$\rho_A(P) = \sum_{k=0}^{m} P_k A^k \text{ (right eval.)}, \qquad \lambda_A(P) = \sum_{k=0}^{m} A^k P_k \text{ (left eval.)}.$$

We derive some basic properties of right evaluation (properties for left evaluation are analogous).

Lemma 5.32 (**Properties of right evaluation**). *Let $A, B \in \mathbf{M}_n(\mathbb{F})$, $P, Q \in \mathbf{M}_n(\mathbb{F}[x])$ and $\alpha, \beta \in \mathbf{M}_n(\mathbb{F})$. Then the right evaluation functon ρ_A satisfies*

$$\rho_A(\alpha P + \beta Q) = \alpha\rho_A(P) + \beta\rho_A(Q) \;\; (\mathbf{M}_n(\mathbb{F}) - \text{linearity}). \tag{5.33}$$

In general, ρ_A is not multiplicative ($\rho_A(PQ) \neq \rho_A(P)\rho_A(Q)$), but we have

$$\rho_A(PQ) = \sum_k P_k \rho_A(Q) A^k \;\; (\text{quasi} - \text{multiplicative property}). \tag{5.34}$$

And as special cases

$$\rho_A(Qx^k I) = \rho_A(Q)A^k \;\text{ and }\; AB = BA \Rightarrow \rho_A(PB) = \rho_A(P)B. \tag{5.35}$$

Proof. **To prove** 5.33: Let $R = \alpha P + \beta Q$ and let $\sum_k R_k x^k I$ be the canonical presentation of R (note change of notation from 5.29). We use 5.29 (uniqueness of canonical presentation). Note that $\rho_A(R) = \sum_k R_k A^k$ becomes

$$\sum_k (\alpha P_k + \beta Q_k) A^k = \sum_k \alpha P_k A^k + \sum_k \beta Q_k A^k = \alpha \rho_A(P) + \beta \rho_A(Q).$$

We take the range of values for k to include all nonzero values of R_k, P_k and Q_k. This proves 5.33.

To prove 5.34: Note that $PQ = (\sum_k P_k x^k I) Q = \sum_k P_k Q x^k I$. Linearity 5.33 (where the $P_k \in \mathbf{M}_n(\mathbb{F})$ play the role of the coefficients $\alpha, \beta \ldots \in \mathbf{M}_n(\mathbb{F})$) implies

$$\rho_A(PQ) = \rho_A\left(\sum_k P_k (Q x^k I)\right) = \sum_k P_k \rho_A(Q x^k I) = \sum_k P_k \rho_A(Q) A^k.$$

The last equality follows from

$$(5.36) \qquad \rho_A(Q x^k I) = \sum_t Q_t A^{k+t} = \left(\sum_t Q_t A^t\right) A^k = \rho_A(Q) A^k$$

which, incidentally, proves the first identity of 5.35. To prove the second identity of 5.35, take $Q = B$ in 5.34, note that $\rho_A(B) = B$ and use $AB = BA$ which implies $BA^k = A^k B$ for all k. This completes the proof of the lemma.

□

Definition 5.37 (**Characteristic matrix and characteristic polynomial**). Let $A \in \mathbf{M}_n(\mathbb{F})$, and let $I \in \mathbf{M}_n(\mathbb{F})$ be the identity. The matrix $xI - A \in \mathbf{M}_n(\mathbb{F}[x])$ is called the *characteristic matrix* of A. The polynomial $\det(xI - A)$ is called the *characteristic polynomial* of A.

Theorem 5.38 (**Equivalence implies similarity**). *Let $A, B \in \mathbf{M}_n(\mathbb{F})$. There exists $P, Q \in \mathrm{GL}_n(\mathbb{F}[x])$ such that $xI - A = P(xI - B)Q$ if and only if there exists $S \in \mathrm{GL}_n(\mathbb{F})$ such that $A = S^{-1}BS$. In words, the characteristic matrices of A and B are equivalent if and only if A and B are similar (4.78). In fact, $S = \rho_A(Q)$ and $S^{-1} = \rho_B(Q^{-1})$.*

Proof. Note that if $A = S^{-1}BS$ then $xI - A = xI - S^{-1}BS = S^{-1}(xI - B)S$. Thus, we assume $P^{-1}(xI - A) = (xI - B)Q$ or $P^{-1}xI - P^{-1}A = QxI - BQ$. Apply ρ_A to both sides and use linearity (5.33):

$$(5.39) \qquad \rho_A(P^{-1}xI) - \rho_A(P^{-1}A) = \rho_A(QxI) - \rho_A(BQ).$$

From 5.35 (first identity) $\rho_A(P^{-1}xI) = \rho_A(P^{-1})A$. From 5.35 (second identity, $B = A$) $\rho_A(P^{-1}A) = \rho_A(P^{-1})A$. From 5.35 (first identity), $\rho_A(QxI) = \rho_A(Q)A$. From 5.33 ($\mathbf{M}_n(\mathbb{F})$- linearity), $\rho_A(BQ) = B\rho_A(Q)$. Substituting these identities into 5.39, we get $\rho_A(Q)A - B\rho_A(Q) = \Theta_n$ or $\rho_A(Q)A = B\rho_A(Q)$. To complete the proof we show that $\rho_A(Q) \in \mathrm{GL}_n(\mathbb{F})$ so we can take $S = \rho_A(Q)$. Let $R = Q^{-1}$ and use 5.34 (quasi-multiplicative property):

$$\rho_A(I) = I = \rho_A(RQ) = \sum_k R_k \rho_A(Q) A^k. \tag{5.40}$$

Note that $\rho_A(Q)A = B\rho_A(Q)$ implies by induction that $\rho_A(Q)A^k = B^k \rho_A(Q)$. Thus, 5.40 becomes $I = \sum_k R_k B^k \rho_A(Q) = \rho_B(R)\rho_A(Q) = \rho_B(Q^{-1})\rho_A(Q)$. This completes the proof. □

Characteristic matrices and polynomials

Theorem 5.41 (**Cayley-Hamilton theorem**). *Let $f_n(x) = \det(xI - A)$ be the characteristic polynomial (5.37) of $A \in \mathbf{M}_n(\mathbb{F})$, $\mathbb{F}$ a field. Then $f_n(A) = \Theta_n$. Alternatively stated, $f_n(A) = \Theta_n$ where f_n is the determinantal divisor of $xI - A$ of highest degree.*

Proof. From definition 3.83 and corollary 3.84

$$\mathrm{adj}(xI - A)\,(xI - A) = I \det(xI - A) = f_n(x)I \tag{5.42}$$

Let $f_n(x)I = \sum_{k=0}^{n} c_k I x^k$. Let $P = \mathrm{adj}(xI - A)$, $Q = (xI - A)$ and apply equation 5.34 of lemma 5.32 (quasi-multiplicative property):

$$f_n(A) = \rho_A(f_n(x)I) = \rho_A(PQ) = \sum_k P_k \rho_A(Q) A^k \tag{5.43}$$

where $\sum_k P_k x^k$, $P_k \in \mathbf{M}_n(\mathbb{F})$, $0 \le k \le n$, is the canonical presentation of P. Substituting $\rho_A(Q) = A - A = \Theta_n$ into 5.43 completes the proof.

□

Remark 5.44 (**Reduced Cayley-Hamilton: $q_n(A) = \Theta_n$**). The Cayley-Hamilton theorem (5.41) states that $f_n(A) = \Theta_n$ where f_n is the determinantal divisor of highest degree of $xI - A$ (i.e., $f_n(x) = \det(xI - A)$ is the characteristic polynomial of A). In fact, $q_n(A) = \Theta_n$ where $q_n(x)$ is the invariant factor of highest degree of $xI - A$. We refer to theorem 5.41, equation 5.42.

$$\mathrm{adj}(xI - A)\,(xI - A) = f_n(x)I. \tag{5.45}$$

By definition, $f_n = q_n f_{n-1}$ so we have

$$\mathrm{adj}(xI - A)\,(xI - A) = q_n(x) f_{n-1}(x) I \tag{5.46}$$

Let $P = \mathrm{adj}(xI - A)$ and note (3.83) that the set of all entries of P is the set of all signed cofactors of $(xI - A)$ and, by definition, f_{n-1} is a greatest common divisor of these cofactors. Let $\hat{P} = P/f_{n-1}$ so that 5.46 becomes

$$\hat{P}\,(xI - A) = q_n(x)I \tag{5.47}$$

where $\gcd\{\hat{P}(i,j) \mid (i,j) \in \underline{n} \times \underline{n}\} = 1$ (up to units). Set $Q = xI - A$ so that, analogous to 5.43, appling equation 5.34 of lemma 5.32 (quasi-multiplicative property) we obtain

$$q_n(A) = \rho_A(q_n(x)I) = \rho_A(\hat{P}Q) = \sum_k \hat{P}_k \rho_A(Q) A^k. \tag{5.48}$$

The fact that $\rho_A(Q) = \rho_A(xI - A) = A - A = \Theta_n$ completes the proof.

Definition 5.49 (**Minimal polynomial**). Let $A \in \mathbf{M}_n(\mathbb{F})$. A polynomial $\phi(x) \in \mathbb{F}[x]$ is called a *minimal polynomial for A* if it has minimal degree among all polynomials $p(x) \in \mathbb{F}[x]$ such that $p(A) = \Theta_{n,n} \equiv \Theta_n$. We call ϕ *the minimal polynomial* if it is monic (i.e., in the canonical SDR for associates for $\mathbb{F}[x]$ 4.36).

Remark 5.50 (**Minimal polynomial and divisibility**). Let $A \in \mathbf{M}_n(\mathbb{F})$. Let $p(x) \in \mathbb{F}[x]$. It is easily seen that if $\phi(x)$ is minimal for A and $p(A) = \Theta_n$ then $\phi(x)$ divides $p(x)$. Otherwise, the remainder $r(x)$ from dividing $p(x)$ by $\phi(x)$ would satisfy $r(A) = \Theta_n$ and would contradict the minimality of $\phi(x)$. Thus, $\phi(x)$ is minimal for A if and only if it divides all $p(x)$ such that $p(A) = \Theta_n$.

Lemma 5.51 (**Minimal polynomial and invariant factors**). *Let $A \in \mathbf{M}_n(\mathbb{F})$ and let $\phi(x) \in \mathbb{F}[x]$ be the minimal polynomial of A. Then $\phi(x) = q_n(x)$ where q_n is the monic invariant factor of $xI - A$ of highest degree.*

Proof. From remarks 5.44 and 5.50 we have that $\phi(x) \mid q_n(x)$. We will show that $q_n(x) \mid \phi(x)$. Consider $\phi(x) - \phi(y) \in \mathbb{F}[x,y]$. Note that $(x-y) \mid (\phi(x) - \phi(y))$ (true for any polynomial $\phi(x)$). Define $\Phi(x,y) \in \mathbb{F}[x,y]$ by $\phi(x) - \phi(y) = (x-y)\Phi(x,y)$. Substituting $x = xI_n$ and $y = A$ preserves this identity since xI_n and A commute. Thus, $\phi(xI_n) - \phi(A) = (xI_n - A)\Phi(xI_n, A)$. Since $\phi(A) = \Theta_n$, we get $\phi(xI_n) = \phi(x)I_n = (xI_n - A)\Phi(xI_n, A)$. Multiply $\phi(x)I_n = (xI_n - A)\Phi(xI_n, A)$ by $\hat{P}$ from equation 5.47.

$$\hat{P}\phi(x)I_n = \hat{P}(xI_n - A)\Phi(xI_n, A) = q_n(x)\Phi(xI_n, A) \tag{5.52}$$

where the entries of $\hat{P}$ are relatively prime. Thus, $q_n \mid \phi(x)$ which was to be shown. □

Rational and Jordan canonical forms.

Definition 5.53 (**Companion matrix**). Let $a = (a_0, \ldots, a_{k-1}) \in \mathbb{F}^k$ where $\mathbb{F}$ is a field of characteristic zero, and let $p_a(x) = x^k - \sum_{j=0}^{k-1} a_j x^j$. Define

$$C(p_a(x)) = \begin{bmatrix} 0 & 1 & 0 & \cdots & 0 & 0 \\ 0 & 0 & 1 & \cdots & 0 & 0 \\ \vdots & \vdots & \vdots & \vdots & \vdots & \vdots \\ 0 & 0 & 0 & \cdots & 1 & 0 \\ 0 & 0 & 0 & \cdots & 0 & 1 \\ a_0 & a_1 & a_2 & \cdots & a_{k-2} & a_{k-1} \end{bmatrix} \tag{5.54}$$

to be the *companion matrix* of $p_a(x)$. If $a = (a_0)$ then $C(p_a(x)) = (a_0)$.

LEMMA 5.55 (**CHARACTERISTIC POLYNOMIAL OF COMPANION MATRIX**). *Let $C = C(p_a(x)) \in \mathbf{M}_k(\mathbb{F})$ be the companion matrix (5.53) of $p_a(x) = x^k - \sum_{j=0}^{k-1} a_j x^j$. Then $\det(xI - C) = p_a(x)$ and the sequence of nonzero determinantal divisors of $D = xI - C$ is $(f_0^D, f_1^D \ldots, f_k^D) = (1, 1, \ldots, 1, p_a)$.*

PROOF. First we show that $\det(xI - C) = p_a(x)$. The proof is by induction. The cases $k = 1, 2$ are easily checked. Assume that the lemma is true for $k - 1$ where $k > 2$. Note that

$$D = xI - C = \begin{bmatrix} x & -1 & 0 & \cdots & 0 & 0 \\ 0 & x & -1 & \cdots & 0 & 0 \\ \vdots & \vdots & \vdots & \vdots & \vdots & \vdots \\ 0 & 0 & 0 & \cdots & -1 & 0 \\ 0 & 0 & 0 & \cdots & x & -1 \\ -a_0 & -a_1 & -a_2 & \cdots & -a_{k-2} & x - a_{k-1} \end{bmatrix}. \tag{5.56}$$

Expanding $\det(xI - C)$ by the first column (Laplace expansion) we get

$$\det(xI - C) = x \det[(xI - C)(1 \mid 1)] + (-a_0)(-1)^{k+1} \det[(xI - C)(k \mid 1)]. \tag{5.57}$$

Note that by the induction hypothesis, $\det[(xI - C)(1 \mid 1)] = p_{a'}(x)$ where $a' = (a_1, \ldots, a_{k-1})$. Note also that $\det[(xI - C)(k \mid 1)] = (-1)^{k-1}$ and thus $(-a_0)(-1)^{k+1} \det[(xI - C)(k \mid 1)] = -a_0$. Substituting these results into 5.57 proves $\det(xI - C) = p_a(x)$. It is easily seen in general that f_{k-1}^D is a unit (consider the submatrix $D(k \mid 1)$).

□

LEMMA 5.58 (**SMITH FORM OF A CHARACTERISTIC MATRIX**). *Let $Q = xI - A$ be the characteristic matrix of $A \in \mathbf{M}_n(\mathbb{F})$. Let $S = \operatorname{diag}(1, \ldots, 1, q_{k+1}, q_{k+2}, \ldots, q_n)$ be a Smith form (4.91, 4.98) of Q where the q_{k+j}, $j = 1, \ldots, n - k$, are the non-unit invariant factors. Then S is equivalent to the direct sum*

$$T = \oplus_{j=1}^{n-k} \operatorname{diag}(1, \ldots, 1, q_{k+j}) \tag{5.59}$$

where $\operatorname{diag}(1, \ldots, 1, q_{k+j}) \in \mathbf{M}_{d_{k+j}}(\mathbb{F}[x])$, $d_{k+j} = \deg(q_{k+j})$, $j = 1, \ldots, n - k$.

PROOF. We show it is possible to rearrange the diagonal entries of S to get T. The theorem requires that $\operatorname{diag}(1, \ldots, 1, q_{k+j})$ has $d_{k+j} - 1$ entries equal to 1. Since $\det(S) = \prod_{j=1}^{n-k} q_{k+j}$ we have $\sum_{j=1}^{n-k} d_{k+j} = n$, and hence $\sum_{j=1}^{n-k}(d_{k+j} - 1) = n - (n - k) = k$ where k is the number of diagonal elements equal to 1 in S. Thus, T is a possible diagonal rearrangement of S. Such a rearrangement of S can be achieved by elementary row and column operations and is equivalent to S. □

DEFINITION 5.60 (**SIMILARITY INVARIANTS**). The invariant factors of the characteristic matrix $xI - A$ of $A \in \mathbf{M}_n(\mathbb{F})$ are called the *similarity invariants* of A.

Lemma 5.61 (Companion matrices of non-unit similarity invariants). *Let $Q = xI - A$ be the characteristic matrix of $A \in \mathbf{M}_n(\mathbb{F})$. Let*

$$S = \text{diag}(1, \ldots, 1, q_{k+1}, q_{k+2}, \ldots, q_n)$$

be a Smith form of Q where the q_{k+j}, $j = 1, \ldots, n-k$, are the non-unit similarity invariants of A.. Then A is similar to $\oplus_{j=1}^{n-k} C(q_{k+j})$ where $C(q_{k+j})$ denotes the companion matrix (5.53) of the similarity invariant q_{k+j}.

Proof. We use theorem 5.38 and lemma 5.55. Each term $\text{diag}(1, \ldots, 1, q_{k+j})$ in equation 5.59 is equivalent to the matrix $xI_{d_{k+j}} - C(q_{k+j})$ where $d_{k+j} = \deg(q_{k+j})$. From 5.59, we have $Q = xI_n - A$ is equivalent to

$$\oplus_{j=1}^{n-k}(xI_{d_{k+j}} - C(q_{k+j})) = I_n x - \oplus_{j=1}^{n-k} C(q_{k+j}).$$

By 5.38, we have A is similar to $\oplus_{j=1}^{n-k} C(q_{k+j})$.

□

Lemma 5.62 (Companion matrices of elementary divisors). *We use the notation of lemma 5.58 and equation 5.59. Let $D_t = \text{diag}(1, \ldots, 1, q_t)$, $k+1 \leq t \leq n$, q_t a non-unit invariant factor of $Q = xI - A$, where $\text{diag}(1, \ldots, 1, q_{k+j}) \in \mathbf{M}_{d_{k+j}}(\mathbb{F}[x])$, $d_{k+j} = \deg(q_{k+j})$, $j = 1, \ldots, n-k$. Let $q_t = p_1^{e_{t1}} p_2^{e_{t2}} \cdots p_s^{e_{ts}}$ where we assume that $e_{tr} > 0$ for $1 \leq r \leq s$ so all $p_r^{e_{tr}}$ are elementary divisors of Q (5.16). Then D_t is equivalent to*

$$\oplus_{r=1}^{s} \text{diag}(1, \ldots, 1, p_r^{e_{tr}}) \tag{5.63}$$

where $\text{diag}(1, \ldots, 1, p_r^{e_{tr}})$ is a $\deg(p_r^{e_{tr}})$ square matrix. Thus, $C(q_t)$ is similar to $\oplus_{r=1}^{s} C(p_r^{e_{tr}})$.

Proof. We have

$$D_t = \text{diag}(1, \ldots, 1, q_t) = \text{diag}(1, \ldots, 1, p_1^{e_{t1}} p_2^{e_{t2}} \cdots p_s^{e_{ts}}).$$

Corollary 4.99, a general statement about Smith forms, implies that D_t is a Smith form of

$$\tilde{D}_t = \text{diag}(1, \ldots, 1, p_1^{e_{t1}}, p_2^{e_{t2}}, \ldots, p_s^{e_{ts}}).$$

The hypothesis of corollary 4.99 that the $p_1^{e_{t1}}, p_2^{e_{t2}}, \ldots, p_s^{e_{ts}}$ are pairwise relatively prime is valid. Both $\tilde{D}_t$ and $D_t \in \mathbf{M}_{n_t}(\mathbb{F})$ where $n_t = \deg(q_t)$. We can now apply the same "rearranging diagonal entries" idea used in the proof of lemma 5.58 to show that $\tilde{D}_t$, and hence D_t, is equivalent to 5.63. Analogous to lemma 5.61, we have that $C(q_t)$ is similar to $\oplus_{r=1}^{s} C(p_r^{e_{tr}})$. This completes the proof.

□

Theorem 5.64 (Frobenius or rational canonical form). *Let $xI - A$ be the characteristic matrix of $A \in \mathrm{M}_n(\mathbb{F})$. Let $\mathcal{G}$ be the multiset of all non-unit elementary divisors of $xI - A$. For $g \in \mathcal{G}$ let $\gamma = \deg(g)$. There exists $P, Q \in \mathrm{GL}_n(\mathbb{F}[x])$ and $S \in \mathrm{GL}_n(F)$ such that*

$$P(xI - A)Q = \bigoplus_{g \in \mathcal{G}} \mathrm{diag}(1, \ldots, 1, g) \tag{5.65}$$

and

$$S^{-1}AS = \bigoplus_{g \in \mathcal{G}} C(g). \tag{5.66}$$

Proof. The proof follows from lemmas 5.55, 5.58, 5.61, and 5.62. □

Definition 5.67 (Hypercompanion matrix). Let $\alpha \in \mathbb{F}$ and $p_\alpha(x) = (x - \alpha)^k$. Define the *hypercompanion matrix* of $p_\alpha(x)$ by

$$H(p_\alpha(x)) = \begin{bmatrix} \alpha & 1 & 0 & \cdots & 0 & 0 \\ 0 & \alpha & 1 & \cdots & 0 & 0 \\ \vdots & \vdots & \vdots & \vdots & \vdots & \vdots \\ 0 & 0 & 0 & \cdots & \alpha & 1 \\ 0 & 0 & 0 & \cdots & 0 & \alpha \end{bmatrix}. \tag{5.68}$$

If $k = 1$ then $H(p_\alpha(x)) = (\alpha)$.

Remark 5.69 (Similarity of companion, hypercompanion matrices). Let $p_a(x) = x^k - \sum_{j=0}^{k-1} a_j x^j$ where $a = (a_0, \ldots, a_{k-1})$ and for $j = 0, \ldots, k$, $-a_j - \binom{k}{j}(-\alpha)^{k-j}x^j$. Then, by the binomial theorem, $p_a(x) = (x - \alpha)^k$. By lemma 5.55, $\det(xI - C(p_\alpha)) = p_a(x) \equiv (x - \alpha)^k$ and the sequence of nonzero determinantal divisors of $D = xI - C(p_\alpha)$ is $(f_0^D, f_1^D \ldots, f_k^D) = (1, 1, \ldots, 1, p_\alpha)$. Note that

$$xI - H(p_\alpha) = \begin{bmatrix} x - \alpha & -1 & 0 & \cdots & 0 & 0 \\ 0 & x - \alpha & -1 & \cdots & 0 & 0 \\ \vdots & \vdots & \vdots & \vdots & \vdots & \vdots \\ 0 & 0 & 0 & \cdots & -1 & 0 \\ 0 & 0 & 0 & \cdots & x - \alpha & -1 \\ 0 & 0 & 0 & \cdots & 0 & x - \alpha \end{bmatrix}. \tag{5.70}$$

It is easily seen that $\hat{D} = xI - H(p_\alpha)$ also has its sequence of nonzero determinantal divisors $(f_0^{\hat{D}}, f_1^{\hat{D}}, \ldots, f_k^{\hat{D}}) = (1, 1, \ldots, 1, p_\alpha)$. Thus, the companion matrix $C(p_\alpha)$ and the hypercompanion matrix $H(p_\alpha)$ have the same similarity invariants and are, therefore, similar matrices (5.38).

THEOREM 5.71 (**JORDAN CANONICAL FORM**). *Let $A \in \mathbf{M}_n(\mathbb{F})$. Let $\mathcal{G}$ be the multiset of elementary divisors of $xI - A$. Assume that every $g \in \mathcal{G}$ is of the form $g = (x - \alpha)^e$, $e > 0$. Then there exists $S \in \mathrm{GL}_n(F)$ such that*

$$S^{-1}AS = \bigoplus_{g \in \mathcal{G}} H(g) \tag{5.72}$$

where $H(g)$ is the hypercompanion matrix of g (5.67).

PROOF. The proof follows from theorem 5.64 (5.66) and remark 5.69. □

REMARK 5.73 (**ELEMENTARY DIVISORS OF DIRECT SUM**). Let $A = B \oplus C$ where $B \in \mathbf{M}_b(\mathbb{F})$ and $C \in \mathbf{M}_c(\mathbb{F})$ and thus $A \in \mathbf{M}_a(\mathbb{F})$ where $a = b + c$. Let $\mathcal{G}^B$ and $\mathcal{G}^C$ be the multisets of elementary divisors of $xI_b - B$ and $xI_c - C$. Let $\mathcal{G} = \mathcal{G}^B \cup \mathcal{G}^C$ be the multiset union of $\mathcal{G}^B$ and $\mathcal{G}^C$ (1.32). From theorem 5.64

$$P_b(xI_b - B)Q_b = \bigoplus_{g \in \mathcal{G}^B} \mathrm{diag}(1, \ldots, 1, g) \tag{5.74}$$

and

$$P_c(xI_c - C)Q_c = \bigoplus_{g \in \mathcal{G}^C} \mathrm{diag}(1, \ldots, 1, g) \tag{5.75}$$

where $P_b,\ Q_b \in \mathrm{GL}_b(\mathbb{F}[x])$ and $P_c\ Q_c \in \mathrm{GL}_c(\mathbb{F}[x])$. We need to prove that, in fact, $\mathcal{G}^A = \mathcal{G}$ where $\mathcal{G}^A$ is the multiset of elementary divisors of $xI_a - A$.

It is clear from equations 5.74 and 5.75 that

$$P_a(xI_a - A)Q_a = \bigoplus_{g \in \mathcal{G}} \mathrm{diag}(1, \ldots, 1, g) \tag{5.76}$$

where $P_a = P_b \bigoplus P_c$ and $Q_a = Q_b \bigoplus Q_c$.

In remark 5.18 we saw how to go back and forth between the multiset of elementary divisors and the non-unit invariant factors if we know the rank. Apply this procedure to the multiset $\mathcal{G}$ knowing a, the rank of A, to obtain the associated (with $\mathcal{G}$) non-unit "invariant factors" $q_t = p_1^{e_{t1}} p_2^{e_{t2}} \cdots p_s^{e_{ts}}$. At this point, we don't know that $\mathcal{G}$ is the list of elementary divisors of $(xI_a - A)$ so we don't know that this list of "invariant factors" is the correct one for $xI_a - A$. In lemma 5.62 we noted the equivalence of

$$D_t = \mathrm{diag}(1, \ldots, 1, q_t) = \mathrm{diag}(1, \ldots, 1, p_1^{e_{t1}} p_2^{e_{t2}} \cdots p_s^{e_{ts}}).$$

and

$$\tilde{D}_t = \mathrm{diag}(1, \ldots, 1, p_1^{e_{t1}}, p_2^{e_{t2}}, \ldots, p_s^{e_{ts}})$$

by using corollary 4.99, noting that the hypothesis of that corollary, that the $p_i^{e_{ti}}$, $i = 1, \ldots s$, are pairwise relatively prime, is valid. Note that the diagonal matrix of 5.76 can be rearranged using row and column matrices so that the $\tilde{D}_t = \mathrm{diag}(1, \ldots, 1, p_1^{e_{t1}}, p_2^{e_{t2}}, \ldots, p_s^{e_{ts}})$ are grouped together. Using row and column operations, convert these $\tilde{D}_t$ to the equivalent D_t. This shows that, in fact, the q_t are the invariant factors of $xI_a - A$. Thus, $\mathcal{G}^A = \mathcal{G}$.

Index

NOTES

www.ingramcontent.com/pod-product-compliance
Lightning Source LLC
Chambersburg PA
CBHW081503170526
45166CB00008B/2534
* 9 7 8 1 5 0 0 2 8 9 5 0 8 *